A Concise History of Radiation Detectors

Hunting for the Unknown

A Concise History of Radiation Detectors

Hunting for the Unknown

Fabio Sauli
CERN, Switzerland

World Scientific

NEW JERSEY • LONDON • SINGAPORE • BEIJING • SHANGHAI • HONG KONG • TAIPEI • CHENNAI • TOKYO

Published by

World Scientific Publishing Co. Pte. Ltd.
5 Toh Tuck Link, Singapore 596224
USA office: 27 Warren Street, Suite 401-402, Hackensack, NJ 07601
UK office: 57 Shelton Street, Covent Garden, London WC2H 9HE

British Library Cataloguing-in-Publication Data
A catalogue record for this book is available from the British Library.

The cover is an iconic illustration of the radiation detectors development over more than a century. The upper image shows a cosmic ray shower, recorded with a photographic emulsion (Powell et al., 1960); cosmic rays were discovered by Victor Hess in 1912, for which he was awarded the 1936 Nobel Prize in Physics. The lower image is the computer rendering of a candidate massive scalar boson, named Higgs from one of the physicists who, in 1964, proposed the theory which led to its prediction; its discovery was announced in 2012 by the experiments ATLAS and CMS at CERN, the European Laboratory for Particle Physics near Geneva, as an outcome of multi-TeV proton-proton collisions at the Large Hadron Collider (LHC). The image is an event recorded with the CMS detector, showing features expected from the decay of the boson onto a pair of photons represented by the dashed yellow lines and green towers; the length of the tracks is descriptive of the particle's energy. Peter Higgs and François Englert were awarded the 2013 Nobel Prize for their theoretical predictions. (© 2012 CERN, for the benefit of the CMS collaboration).

A CONCISE HISTORY OF RADIATION DETECTORS
Hunting for the Unknown

ISBN 978-981-98-1939-3 (hardcover)
ISBN 978-981-98-1940-9 (ebook for institutions)
ISBN 978-981-98-1941-6 (ebook for individuals)

For any available supplementary material, please visit
https://www.worldscientific.com/worldscibooks/10.1142/14485#t=suppl

Desk Editors: Joseph Ang/Rhaimie Wahap

Typeset by Stallion Press
Email: enquiries@stallionpress.com

Preface

The development of radiation detectors has an appealing history, closely entangled with the discovery of elementary particles and of their interactions. Invented in the late eighteen century, the gold leaf electroscope allowed to detect small amounts of electric charges, including those created by ionizing radiation; in 1912, using improved sensitivity electrometers developed by Theodor Wulf, Victor Hess discovered the cosmic rays in a balloon flight experiment. Portable small-size electrometers have been for years in wide use for radiation monitoring; ionization chambers, modern-day versions of the electrometers, are still used to measure the electron and ion charges released in gas-filled vessels and collected on electrodes under the effect of an applied electric field.

Photography, developed in the late 18th century, although not purposely designed for radiation detection, contributed to the discovery of ionizing radiations and to our understanding of their properties. In 1895 Wilhelm Röntgen, investigating the properties of low-pressure electrical discharges, observed the emission of an unexplained radiation that could penetrate opaque materials, whilst darkening a photographic plate; he appropriately named it "X-rays". A year later, Henry Becquerel, inadvertently exposing a stack of photographic emulsions to uranium salts, discovered natural radioactivity; this observation further stimulated the elaboration of more performing radiation detectors. High-sensitivity photographic emulsions have been used for years to study the nature and properties of the cosmic ray flux and of the outcomes of their interactions with the atmosphere; nuclear emulsions contributed to the discovery by Cecil Frank Powell and his team of the pion in 1947 and of the K^- meson in 1950.

Developed in the early 1900s by C.T. Wilson, the cloud chamber permitted to visualize ionizing radiation, creating condensation centres around their trajectories in super-saturated vapours; it led to the discovery by Carl Anderson of the positron in 1932 and of the muon in 1936. The bubble chamber, invented in 1952 by D.A. Glaser, permitted also to visualize with excellent resolution ionizing tracks in a vessel filled with a superheated liquid, and was used for many years in systematic studies of the interactions of high-energy particles produced at accelerator-driven experiments.

Cloud and bubble chambers are sensitive during a short time, unrelated to the source of radiation. Developed by S. Fukui and S. Miyamoto in the late fifties of last century, spark chambers permitted to reveal ionizing radiation under control of an electronic selection, thus privileging events of interest. Streamer chambers extended the visualization of complex events to large detection volumes.

The yields of natural radioactivity, named α, β and γ rays, experience ionizing encounters copious enough to be recorded with emulsions or simple charge-sensitive instruments. Developed in the early 1900 by Ernest Rutherford and Hans Geiger, and perfectioned by Walther Müller, wire gaseous counter multiply the released electron charge by large factors, permitting the detection of small ionization yields, down to single electrons. Providing a charge signal multiple of the primary ionization, proportional counters have been and still are widely used for radiation measurements and monitoring. In their modern embodiments, position-sensitive multiwire and drift chambers, invented by Georges Charpak in 1967, gaseous counters instrument most large detection systems deployed in particle physics and astrophysics, and led to major discoveries such as the intermediate vector bosons W and Z and the Higgs particle. The recently developed wire-less devices, named micropattern gaseous counters, offer superior performances at high radiation fluxes and are gradually replacing wire chambers in experimental setups.

Perfectioned in the fifties of last century, silicon semiconductors opened the way to a novel and powerful line of radiation detectors. Owing to their high density and low ionization energy, as compared to gaseous devices, semiconductors provide large charge signals, relatively easy to reveal with suitable electronic circuitry. Evolving from single crystals to more elaborated multi-electrode structures, silicon microstrips and pad arrays have progressively replaced in particle physics detectors less performing devices close to the interaction region in colliding beam accelerators; the multi-megachannels silicon strips and pads systems of the CMS an ATLAS experiments at CERN's LHC were instrumental for the discovery of the Higgs heavy boson in 2012. Dedicated designs of the previous detectors and their combinations are deployed in the hunt of a host of hypothetical new particles: axions, dark matter, weakly interacting massive particles and more.

This book is an attempt to review more than one century of progress of radiation detectors, discussing examples of their use in a wide range of applications, from particle to nuclear physics, plasma diagnostics, astrophysics, biology, medicine. Emphasis is put on the operating principles of the devices, as well as on their comparative performances.

Most illustrations stem from the perusal of the vast number of articles published on the subject, as well as from web-based search algorithms; the quality of the artwork is consequently often not optimal. As it has not been possible to always ascertain the origin of the documents, the author apologizes in advance for any omission or copyright infringement, and is available to correct any ensuing issue.

Owing to the lifelong association of the author with the European Organisation for Nuclear Research (CERN) based in Geneva, Switzerland, and the easier access to the institution internal documents, preference is often given to describe CERN-based advancements; similar and often improved works realised in other laboratories are illustrated whenever possible.

A non-exhaustive list of major books and review articles related to the subjects covered in this book is given under "Basic Bibliography"; specific references are quoted within the various chapters.

The purpose of the book is to provide a simple yet organic impression of the past and ongoing research on radiation detectors, referring the readers for details on the different technologies to the more specialized works in the quoted literature.

My warmest appreciation for the help of my son-in-law Stephen Winfield, who undertook the arduous task of improving the language on a subject extraneous to his field of competence.

Contents

Chapter 1

Photographic Plate Emulsions

Photographic emulsions, developed in the late nineteen century, record visible light images exploiting the chemical reaction between silver halide grains and photons; the grains are normally embedded in a thin colloidal gelatine layer, named emulsion, and coated on a glass substrate. After exposure to light, the exposed grains are reduced to black metallic silver particles by immersion in a developer solution, fixing the image.

The first observation of sensitivity of photographic plates to non-visible radiation is due to Wilhelm Rôntgen in 1895, investigating the properties of low-pressure gas discharges; the emitted radiation could penetrate thick absorbing materials while darkening a photographic emulsion, and was appropriately named "X". In 1901, Röntgen was awarded the first Nobel Prize in Physics for his discovery. The first X-Ray radiographies of the hand of Röntgen wife are historical, Figure 1.1. A year later Henry Becquerel, accidentally exposing a stack of photographic plates to uranium salts, discovered natural radioactivity, Figure 1.2; for a recent note on the history of the finding see (Sekiya and Yamasaki, 2015). This observation further stimulated the development of performing radiation detectors making use of photographic emulsions.

High-sensitivity photographic film has been used for decades in the medical field to register X-ray radiographies; however, as the conversion efficiency of the film is rather low for hard X-rays, the patient was exposed to relatively large doses. A substantial reduction, up to 90–95%, is obtained with the use of intensifying screens, where the X-rays are

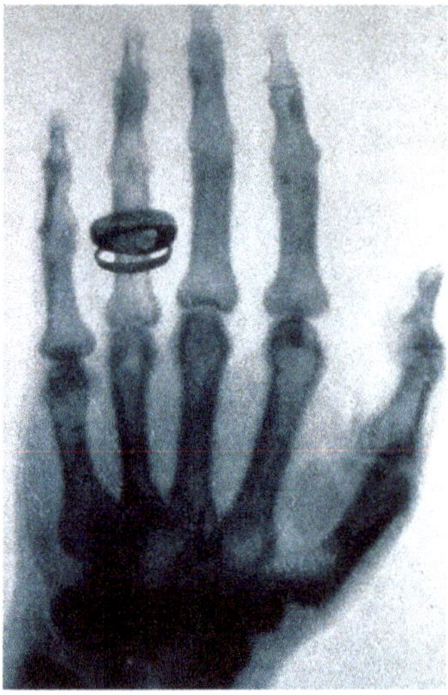

Figure 1.1. Historical X-ray radiography of Röntgen wife's hand.

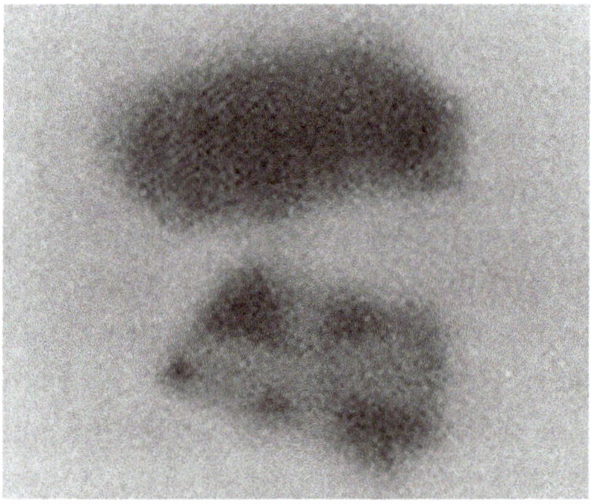

Figure 1.2. Becquerel's discovery of uranium natural radioactivity.

absorbed in a thick phosphor layer emitting visible light that is then recorded on photosensitive plates.

Emulsions were found capable to record ionizing tracks in the early 1900, and used then for decades as one of the major tools to study radiation sources, cosmic rays and radioactive elements. The early works were carried out with the micron-thin emulsions used for conventional photography, but the development in the thirties of thicker layers, 50 μm and more, extended the sensitivity range of the plates. Further improvements in the fifties included a tenfold increase of the density of the silver grains, of the sensitivity and of the thickness permitting to register minimum ionizing particle tracks (Beiser, 1952).

Originally used to provide qualitative information on the radiation, systematic investigations of the emulsion properties permitted to obtain accurate quantitative results of the recorded tracks; for a summary of these advancements see for example Ref. (Herz and Lock, 1966).

While limited in data acquisition rates by the laborious methods of image analysis, usually done under a microscope, nuclear emulsions offer exceptional spatial resolutions; their limited weight allow the plates to be carried at high altitudes and balloon flights. Figure 1.3 is an example of a secondary particles shower generated by an interacting high-energy cosmic ray in the emulsion (R. R. Daniel *et al.*, 1952). The event topology is recorded for all charged tracks contained in the emulsion layer, and the density of the grains provides information on the differential energy loss of the prongs. Figure 1.4 shows the famous observation of the τ meson (later named kaon) in an emulsion exposed to cosmic rays at the Jungfraujoch High Altitude Observatory (Brown *et al.*, 1949).

Large stacks of nuclear emulsions plates were widely used in the early sixties to investigate the interaction yields produced in the initial operation of accelerators at CERN, the European Laboratory for Nuclear Research in Geneva. Combined with the information obtained with other technologies (spark, bubble chambers) it led to the discovery of a new family of short-lived high-mass baryons named hyperons, Λ, Σ and Ω,

Figure 1.3. Secondary particles shower generated by the interaction of a high-energy cosmic ray in an emulsion (R. R. Daniel *et al.*, 1952).

Figure 1.4. Historical observation of the τ meson (later named kaon) in an emulsion exposed to cosmic rays. The τ enters the plate from the top right and interacts emitting two π^+ and one π^- mesons (Brown *et al.*, 1949).

later found to be characterized by a new quantum number named strangeness. Figure 1.5 is the first observation of an "hyperfragment"; induced by a high-energy proton interaction in the emulsion, a nuclear fragment *F* disintegrates at the point *B*, and a reconstruction of the topology of the three-prong event indicates that the fragment was a Λ^0 hyperon (Herz and Lock, 1966).

The development of fully automatized optical scanning system made it possible the use of nuclear emulsions in modern particle physics detectors. The massive, high granularity OPERA neutrino oscillations experiment, for example, is a hybrid apparatus composed of electronic detectors and lead-emulsion structures. An advanced track finding algorithm permits to follow the products of an interaction and map with great accuracy the topology of the neutrino interaction outcomes in the emulsion (Ieva, 2010). Figure 1.6 is an example of a 3D charm-like topology

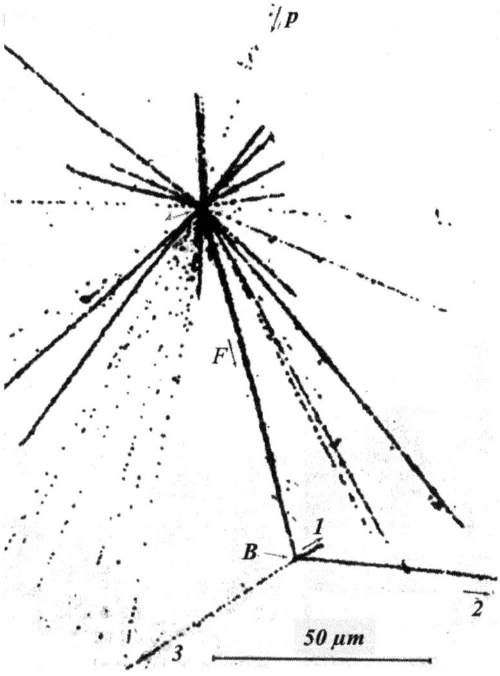

Figure 1.5. A proton-induced hyperfragment (Herz and Lock, 1966).

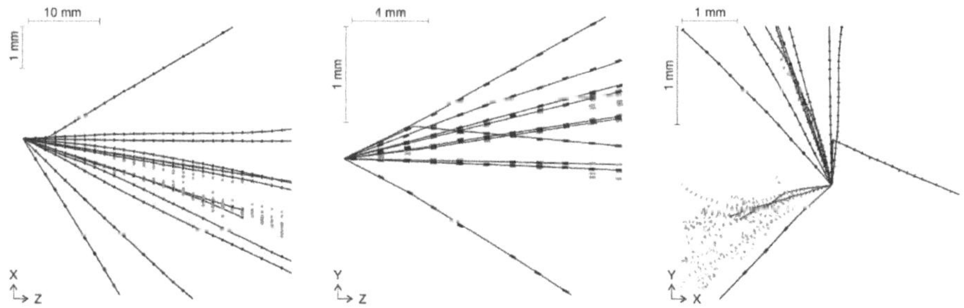

Figure 1.6. Emulsion reconstruction of a neutrino interaction (top, side and front views) (Agafonova *et al.*, 2009).

event as seen in the stack of emulsions following the neutrino interaction point, revealed by the electronic section of the experiment.[1]

[1]In 2011, from the measured time difference between generation at CERN and detection, the experiment incorrectly announced that neutrinos appeared to travel faster than light. The source of error was found a few months later.

Chapter 2

Cloud Chambers

Developed in the early 1900 by Charles Thompson Rees Wilson, the cloud chamber consists in a gas-tight vessel containing a supersaturated vapour of water or alcohol. When charged particles ionize the medium, the released charges act as condensation centres, forming trails of droplets that can be visualized or photographed; the condition of supersaturation is reached with a rapid expansion of the vessel. Figure 2.1 is one of the early images recorded by Wilson with his expansion cloud chamber exposed to alpha particles emitted by a radium sample.

Using an expansion cloud chamber Carl David Anderson discovered the positron in 1932, and the muon in 1936; Figure 2.2 is the historical picture leading to the positron discovery.

In 1936, Alexander Langsdorf devised the diffusion cloud chamber, where the condition of supersaturation is achieved with a temperature gradient between the top and the bottom of the vessel, thus avoiding the need of a mechanical expansion piston. For nearly 30 years, the cloud chamber was one of the major tools for particle physics investigations, progressively replaced by the invention of the bubble and spark chambers. It remains today a popular exhibit in science museums; an example is shown in Figure 2.3.

For his invention, C. T. Wilson received the 1927 Nobel prize for Physics; Anderson obtained the Nobel prize in 1936, shared with Victor Hess for their discoveries.

Originally published in 1951, H. J. Wilson book on the cloud chamber technique has recently been reprinted (Wilson, 2014).

Figure 2.1. Early cloud chamber image of emissions from a radium source.

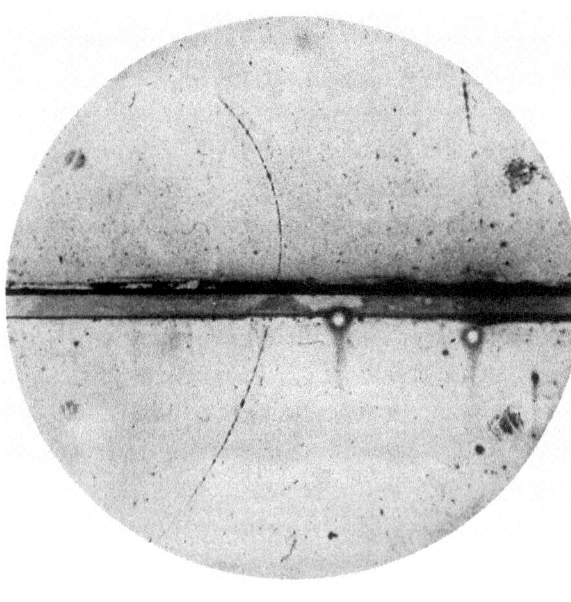

Figure 2.2. Discovery of the positron. A positively charged particle, as revealed by the bending in magnetic field, enters the detector from the bottom, loses energy in a thin lead plate in the centre of the picture and emerges with lower energy with a smaller radius.

Figure 2.3. Cosmic ray shower recorded in a diffusion cloud chamber. A magnetic field creates spiralizing tracks for low-energy delta electrons.

Chapter 3

Bubble Chambers

Invented in 1952 by Donald A. Glaser at the Lawrence Berkeley Laboratory (LBL), the bubble chamber consists in a large cylinder filled with a transparent liquid heated just below its boiling point. In coincidence with charged particles crossing the chamber, a piston in the vessel's top is swiftly upthrust, decreasing the liquid pressure and bringing it to a superheated state. This results in the creation of trails of microscopic bubbles along the particles trajectories, that can be observed and photographed; the density of ionization depends on mass and energy of the charged particles. An applied magnetic field bends the charged particles; analysis of the curvature of the tracks provides their polarity and momentum.

Due to the brief latency of the ionization trails, the volume expansion has to be made within few tens of milliseconds from the event, a rather demanding requirement given the weight of the piston, several tons in large devices; the expansion rate is also limited by the recovery time of the mechanical system. For a discussion of bubble chambers construction, performances and applications see Refs. (Henderson, 1970) and (Bettelli *et al.*, 1993).

Figure 3.1 is an image recorded with the 12" propane-filled bubble chamber built and operated at the Columbia University NEVIS laboratories by Jack Steinberger and collaborators. A 3-D image of the events is obtained using triplets of synchronized cameras.

With the commissioning of novel and powerful particle accelerators, and despite their limits in event rates due to the mechanical expansion

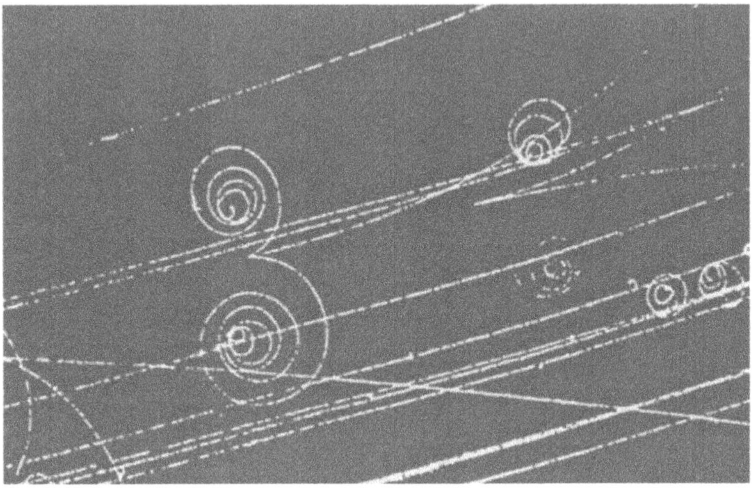

Figure 3.1. Example of tracks recorded with a propane bubble chamber. Gamma-generated electron–positron pairs and delta electrons spiralizing in magnetic field are recognizable.

process and the photographic recording, bubble chambers provided a unique tool for the studies of fundamental particle physics; increasingly large devices were built in major laboratories world-wide. The development of semi-automatic events digitization, replacing the laborious manual film scanning, permitted to study rare events like the new families of strange particles and their resonances. Figures 3.2 and 3.3 show CERN's 2 m liquid hydrogen bubble chamber, built in 1958, and the 35 m^3 Big European Bubble Chamber (BEBC), operated from 1973 to 1984 at CERN with liquid hydrogen, deuterium or neon. In 1973, the neutral currents were discovered with GARGAMELLE, the heavy liquid bubble chamber at CERN, exposed to neutrino beams, providing the first experimental indication of the existence of the Z_0 boson. Figure 3.4 gives an example of raw and reconstructed event, recorded with the British 1.5 m hydrogen bubble chamber at CERN.

Figure 3.2. The 2 m hydrogen bubble chamber at CERN.

Similar devices were operated in other laboratories: Brookhaven National Laboratory (BNL), FERMILAB and Stanford Linear Accelerator (SLAC). Discontinued in the nineties in favour of more performing, fast and fully electronic detectors, bubble chambers have successfully contributed to our basic understanding of particle physics.

Figure 3.3. The BEBC main body, exposed at CERN's open-air museum.

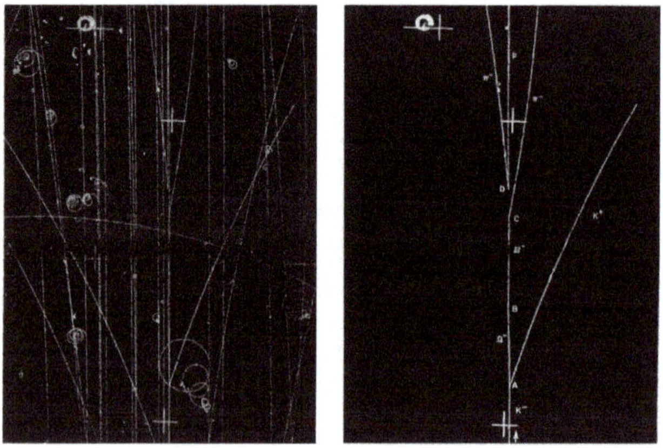

Figure 3.4. Raw and reconstructed K^--p interaction. The sequence of events is: $K^- + p \to K^+ + \Omega^- + K^0$, $\Omega^- \to \Xi^- + \pi^0$, $\Xi^- \to \Lambda + \pi^-$, $\Lambda \to p + \pi^-$. (Picture CERN).

Chapter 4

Spark and Streamer Chambers

4.1 Optical Spark Chambers

Invented in 1959 by Shuji Fukui and Shigenori Miyamoto (Fukui and Miyamoto, 1959), the discharge or spark chamber consists in set of parallel metallic electrodes, alternatively connected to ground and to a high-voltage pulse generator. When a charged particle crosses the structure, as revealed by one or more scintillation counters, an HV pulse is swiftly applied to alternate plates, prompting the formation of sparks along the particle trajectory (Figure 4.1). Figure 4.2 is an image recorded by the inventors of the device.

To be effective, the HV pulse has to be applied within few hundred nanoseconds from the passage of the particle; simple triggered spark gaps, similar to the plugs used for cars, or more sophisticated generators have been developed to this extent, described for example in the book by P. Rice-Evans (1974). A small electric field is usually applied between the plates to clear the gaps from the released charges; this permits to operate the device at high rates, only limited by the recovery time of the pulse generator and of the data recording system.

Figure 4.3 shows a multigap spark chamber stack built at CERN, and Figure 4.4 is the image of a cosmic particle crossing the structure. A medium-size optical spark chambers setup operated at CERN in the late 1960 is shown in Figure 4.5.

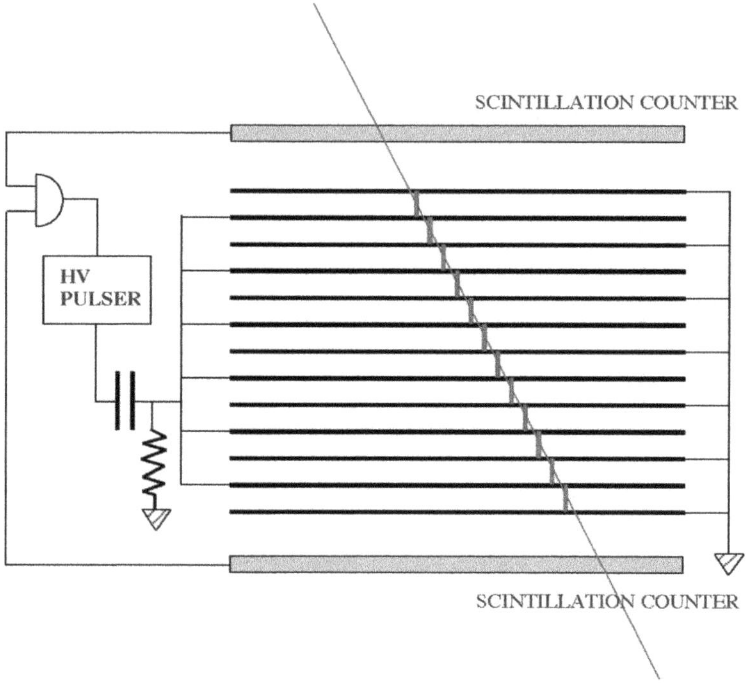

Figure 4.1. The spark chamber: a stack of metallic plates, alternatively connected to ground and to a high-voltage pulser.

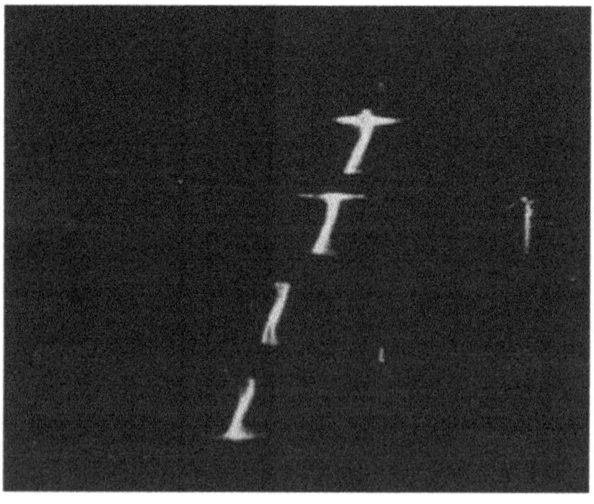

Figure 4.2. Photographic recording of a track in Fukui's original spark chamber.

Figure 4.3. A multigap spark chamber stack. To permit photographic recording of the events, the side frames are made of transparent plastics (Picture CERN).

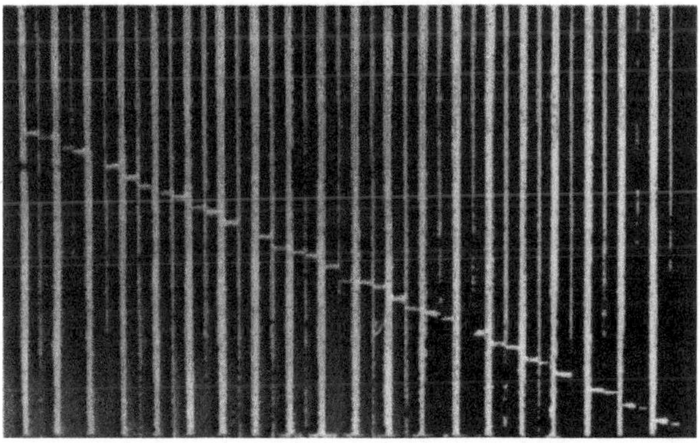

Figure 4.4. A charged particle track recorded with the multigap spark chamber.

Analysis of the photographic recording of events requires a lengthy processing of the films. A significant progress was achieved in the early seventies with the use of the newly developed television tubes, vidicon and plumbicon, scanning and digitizing the images in real time. Figure 4.6 is a curious image of an interaction in the Omega magnetic spectrometer at CERN, recorded by a TV digitizer.

Figure 4.5. The NA11 optical spark chambers setup at CERN (1969).

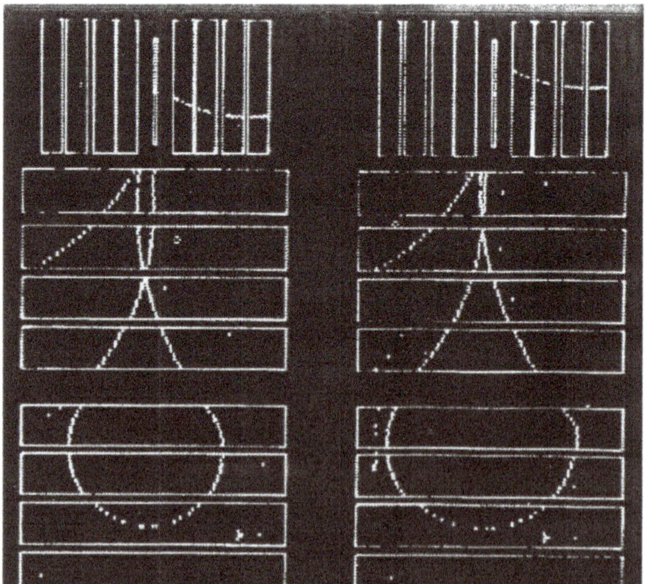

Figure 4.6. A curious event recorded with a television digitizer at CERN's OMEGA spectrometer (1972).

Addition in 1975 to the detector of a large wire drift chamber improved the localization accuracy for forward tracks. The Omega spectrometer played a key role in the discovery of charmonium states, which are bound states of a charm quark and its antiquark.

The advancements of automatic electronic readout systems put an end in the eighties to the utilisation of optical spark chambers in particle physics (Charpak, 1970); they are still used in dedicated cosmic rays experiments and in scientific exhibits in several laboratories.

4.2 Digital Spark Chambers

In the early sixties, a method of digital readout of spark chambers utilizing the know-how acquired in the evolution of computer memories using ferrite cores was proposed by Frank Krienen at CERN (Krienen, 1962). Replacing the planar electrodes, sets of parallel wires are interlaced with miniature ferrite cores, Figure 4.7 (Aubert *et al.*, 1971). The current induced on a wire by a discharge flips the status of the core, that could then be sensed by suitable read pulses under control of routing electronics, following the scheme

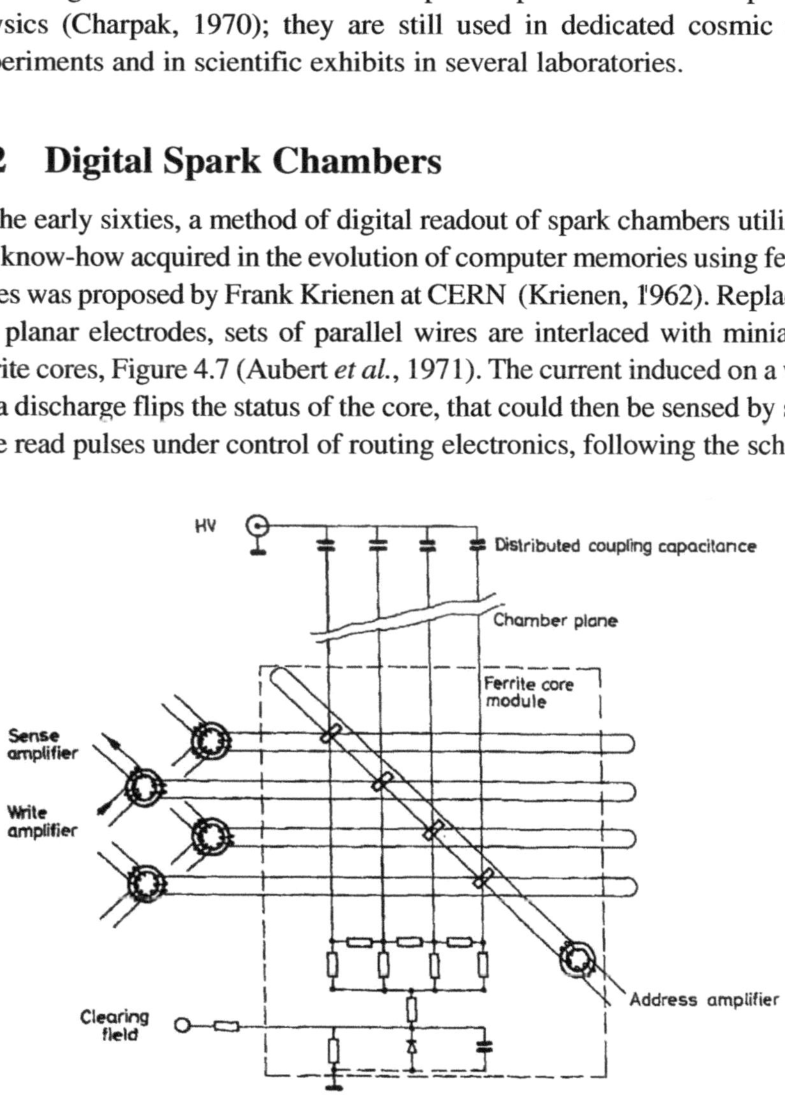

Figure 4.7. Readout of a wire chamber plane with ferrite cores (Aubert *et al.*, 1971).

used for computer memories. At the end of the read cycle, all beads are reset to their quiescent status; typically a full sequence woold take 30–50 μs. Two-dimensional coordinate readout could be achieved using sets of perpendicular wires in the detector; alternatively, several in-line chambers, rotated around the central direction, ensure ambiguity-free reconstruction. With a position accuracy nearing the wire spacing (typically 1 mm) the detector could record events at rates only limited by the readout time and recovery of the high voltage pulsing system.

A large system including more than 30 large size wire spark chambers with digital ferrites readout was deployed at CERN for the study of the time dependence of decay modes of neutral kaons (Aubert *et al.*, 1971); similar setups existed at DESY, FERMILAB, the Joint Institute for Nuclear Research (JINR) in Dubna and other laboratories.

An alternative and elegant method of implementing a digital readout of wire spark chambers was suggested in 1965 by Victor Perez-Mendez, then at Lawrence Berkeley Laboratory. It consists in a wire or ribbon of magnetostrictive material laid perpendicular in contact with the chamber wires; the current pulse flowing in a fired wire couples magnetically to the sensor inducing a local mechanical deformation (striction) that propagates to both ends of the wire, where a signal is detected by pick-up coils. The spark coordinates are determined by digitizing the time delay from the chamber trigger signal to the arrival of the magnetostrictive pulse at the ends of the pick-up wire. Multiple-spark events are recorded without ambiguities; the only limitations are the two-spark resolution of the transducer signal and the readout capability of the digitizing electronics.

Figure 4.8 shows the author of this book with a detector setup based on spark chambers with magnetostrictive readout, built for an experiment at CERN to study ion-deuteron elastic scattering, and operated for several years in the late 1960–early 1970 (F. Bradamante and F. Sauli , 1967).

The introduction in 1967 by Georges Charpak of the multiwire proportional chamber, with performances surpassing by orders of magnitude spark-based detectors, gradually led to abandon these devices.

4.3 Streamer Chambers

In spark chambers, on application of the HV pulse, inelastic electron-molecules collisions generate an avalanche charge multiplication process,

Figure 4.8. F. Sauli with the CERN-Trieste magnetostrictive spark chambers setup (Picture CERN, 1967).

followed by the transition to a streamer and, for thin-gap devices, a full discharge from anode to cathode. The duration of the high-voltage pulse can be truncated to restrict the formation of an avalanche, or protracted to reach a full streamer development.

Developed by Giorgi Chikovani and his group at the Institute of Physics in Tbilissi, and further improved by Boris Dolgoshein and collaborators at the Moskow Lebedev Physical Institute, the streamer chamber consists in a sizeable gas volume between two electrodes; on application of a very brief high-voltage voltage pulse, typically several hundred kV for a few nanoseconds, short visible discharges (avalanches or streamers) develop along the trajectories of ionizing particles and permit to visualize the events (Chikovani *et al.*, 1963; Dolgoshein *et al.*, 1957). Restricting the length of

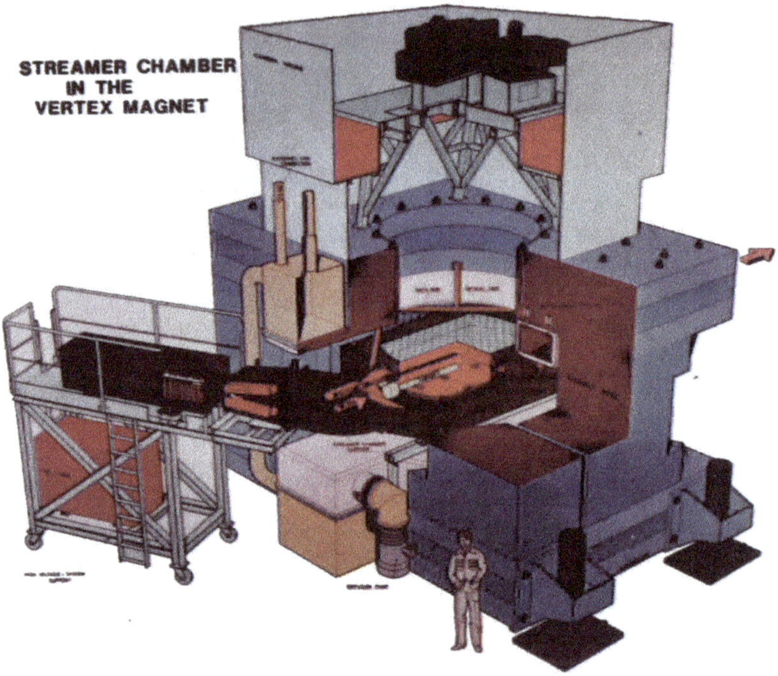

Figure 4.9. Cutaway drawing of the NA35 streamer chamber at CERN (Pühlhofer *et al.*, 1988).

the HV pulse, the detector can be operated in the avalanche mode, before full development of the streamer; while resulting in smaller amount of emitted light, the avalanche mode offers better space resolution. As in the case of spark chambers, events could be recorded using conventional photography or digitized using television cameras.

Usually embedded in a large magnet permitting the momentum analysis of recorded tracks, streamer chambers were rather massive and complex devices. Figure 4.9 shows the NA35 streamer chamber complex, operated in the eighties at CERN. Although rather slow, due to the limitation in generating the large HV pulses, streamer chambers provide a very detailed image of complex events; Figure 4.10 is an example of an event recorded with the UA5 streamer chamber operated in the avalanche mode.

Figure 4.10. First interactions at CERN proton-antiproton collider recorded with the UA5 avalanche chamber (Picture CERN, 1981).

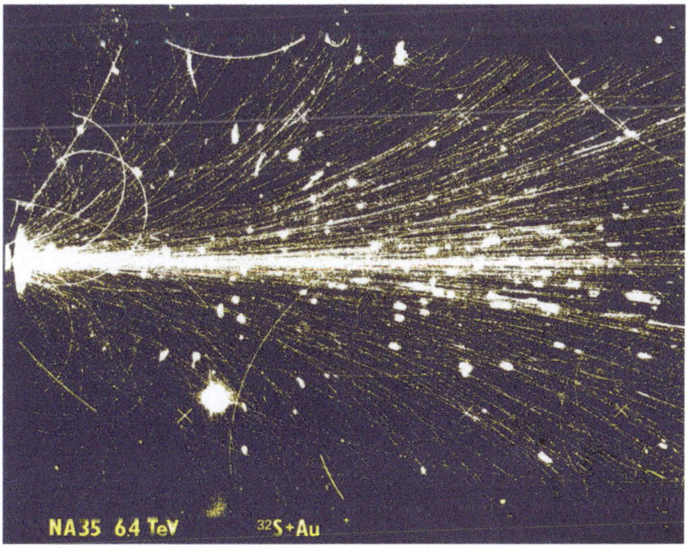

Figure 4.11. A 65 TeV sulphur-gold collision in the NA35 streamer chamber (Picture CERN, 1987).

Figure 4.11 shows an event generated by the collision of very high momentum sulphur ions with a gold target in the ~2 m^3 NA35 streamer chamber at CERN. The experiment focused on studying the relativistic nucleus-nucleus collisions in search of new types of quark matter, and provided the fist evidence of the quark-gluon plasma in 1995. Streamer chambers of similar size were deployed in experiments in other high-energy physics laboratories.

A good summary of evolution and use of streamer chambers at CERN is given in the 1988 review paper by François Rohrbach (1988).

Chapter 5

Scintillators and Photon Detectors

The phosphorescence of materials exposed to X-rays was observed in the late 1800s, in the course of the development of the Crookes discharge tubes. In 1886, accidentally exposing to uranium salts a photographic plate wrapped with powdered zinc sulphate crystals, Henry Becquerel discovered natural radioactivity; in 1903, he shared with Pierre and Marie Curie the Nobel Prize in Physics for the discovery.

Studying the angular distribution of alpha particles scattered by a gold foil, as revealed on a $ZnSO_4$-coated screen, Lord Rutherford developed his theory of the atomic nucleus structure, for which he was awarded the 1908 Nobel Prize in Chemistry.

Scintillators are materials emitting light when excited by radiation; detected by suitable sensors, photodiodes or photomultipliers, scintillation photons provide an electrical signal that can be revealed. An ideal scintillator has a large and linear photon yield in a wide range of energy loss; it should be transparent to its own emission, and have good and uniform optical properties. In many applications, a short decay time of the induced luminescence is paramount for high-rate performances.

Two main processes contribute to the photon emission of a material under irradiation: fluorescence, the prompt emission of visible radiation following its excitation, and phosphorescence, a delayed emission at longer wavelengths. To be a good scintillator for use in particle physics, a material should convert as much as possible of the energy loss in prompt fluorescence.

Over the years, a large variety of scintillating materials has been developed, belonging to two groups: inorganic crystals, and organic plastic. An extended inventory of scintillator material with their main characteristics can be found in the web-based Lawrence Berkeley Laboratory compilation (*Scintillator Library*, 2024); a comprehensive list of commonly used scintillators as well as a summary of their performances can be found in Glenn Knoll's textbook (Knoll, 2010).

5.1 Inorganic Crystal Scintillators

After the initial works with powdered crystals, the use of inorganic scintillators remained limited owing to the difficulty of manufacturing pure and large size crystals. Crystalline sodium iodide (NaI) was a promising material but suffered from poor scintillation yield, mainly due to the large gap between valence and conduction gap. In 1949, Robert Hofstadter at Stanford University discovered that the addition to the crystals of molecules with intermediate energy levels (activators) largely increased the scintillation yield, demonstrated the superior performances in detecting gamma rays of thallium-activated sodium iodide [NaI(Tl)], and developed the technology permitting to manufacture large and uniform crystals. Providing an exceptional energy resolution owing to its scintillation yield close to 40,000 photons/MeV, NaI(Tl) has been widely used in a gamma-ray spectrometers for nuclear and high-energy physics and astrophysics, as well as in medicine, biology, chemistry, geology and other fields. In 1961, Hofstadter shared with Rudolf Mössbauer the Nobel prize for physics for his pioneering studies of electron scattering in atomic nuclei, performed with the gamma ray spectrometer. The scintillation yield of NaI(Tl) has become the reference for other scintillators.

Beginning in the eighties of last century, and continuing today, a search for high-density scintillating crystals has been motivated by the requirements in particle physics and medical diagnostics. Table 5.1 provides the main properties of selected crystal scintillators.

Barium fluoride has a fast scintillation component in the ultraviolet, exploited for fast electromagnetic calorimetry. The high-density, high atomic number bismuth germanate (BGO), lead tungstate (PWO) and lutetium-yttrium orthosilicate crystals (LYSO) are very efficient for

Table 5.1. Main characteristics of selected crystal scintillators, from various sources.

Material	λ_{MAX}(nm)	τ_{DECAY} (μs)	ρ (g/cm^3)	Photons/MeV
NaI(Tl)	415	0.230	3.67	38,000
CsI(Tl)	540	0.68, 3.34	4.51	65,000
BaF_2 (fast)	220	0.0009	4.89	1,400
BaF_2 (slow)	310	0.63	4.89	9,500
BGO ($Bi_4Ge_3O_{12}$)	480	0.34	7.13	8,200
PWO ($PbWO_4$)	440	0.015	8.2	490
LYSO	420	0.047	7.4	25,000

detecting hard X-rays and gamma rays; they are used in large arrays in particle physics calorimeters, nuclear medicine, positron emission tomography (PET). BGO has a large light output, but a slow decay time, while PWO, better suited for high-rate applications, has a rather small yield; it has been adopted to detect high-energy gamma rays by CERN's Compact Muon Solenoid electromagnetic calorimeter (CMS ECAL). The recently developed but expensive LYSO (Lutetium-Yttrium Orthosilicate) is a preferred choice for PET detectors due to its short decay time, permitting a fast coincidence between the two branches of the imager thus improving the signal over noise ratio.

Crystals manufacturing methods have been improved in view of very large production, as for the CMS ECAL (CMS_Collaboration, 1997); Figure 5.1 shows raw PWO ingots before cutting and polishing, and Figure 5.2 a finished crystal. Figure 5.3 is a modular assembly of PWO crystals.

5.2 Organic Scintillators

Unlike inorganic scintillators, necessitating a regular crystalline lattice to be effective, for some organic compound the fluorescence mechanism occurs at the molecular level and is therefore independent from their physical states. Particularly efficient organic scintillators are aromatic hydrocarbons and their derivates: benzene, anthracene and stilbene. The

Figure 5.1. Raw PWO ingots.

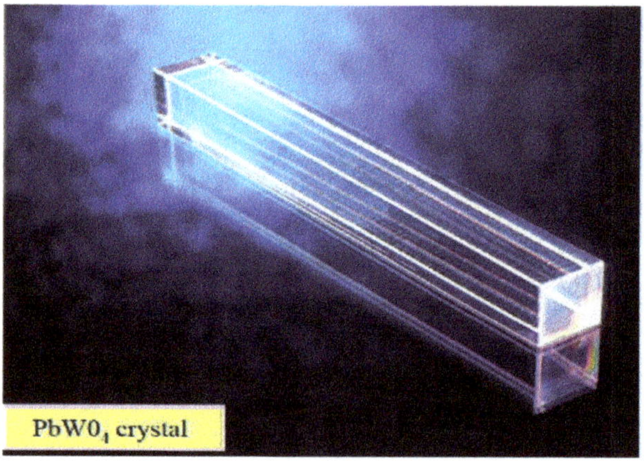

Figure 5.2. Cut and polished crystal.

scintillator can be dissolved in a solvent and then polymerized to form a solid plastic block that can be moulded into the desired shapes; commonly used solvents are styrene, polyvinyl toluene and polystyrene.

Scintillators are characterized by an optical absorption and an emission spectrum at longer wavelengths, as shown in Figure 5.4; to be

Figure 5.3. A CMS ECAL module (Picture CERN).

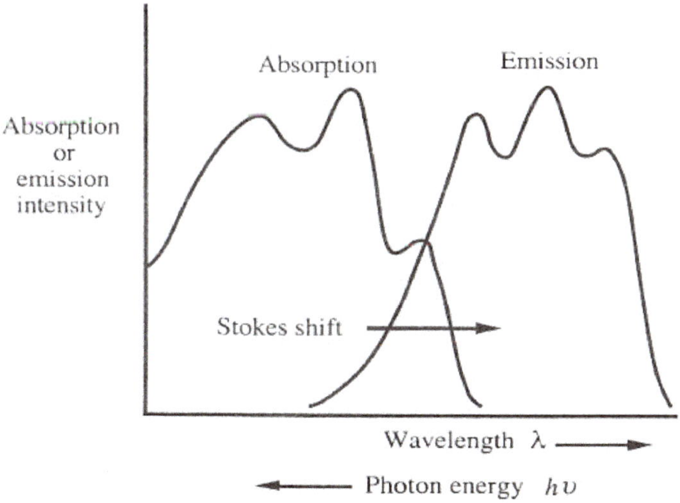

Figure 5.4. Example of optical absorption and emission spectra.

valuable as detector, the material should be transparent to the emission spectrum. This is achieved adding a wavelength shifter (WLS), a compound with the absorption spectrum overlapping with the emission of the main material. The process can be reiterated with two or more additives,

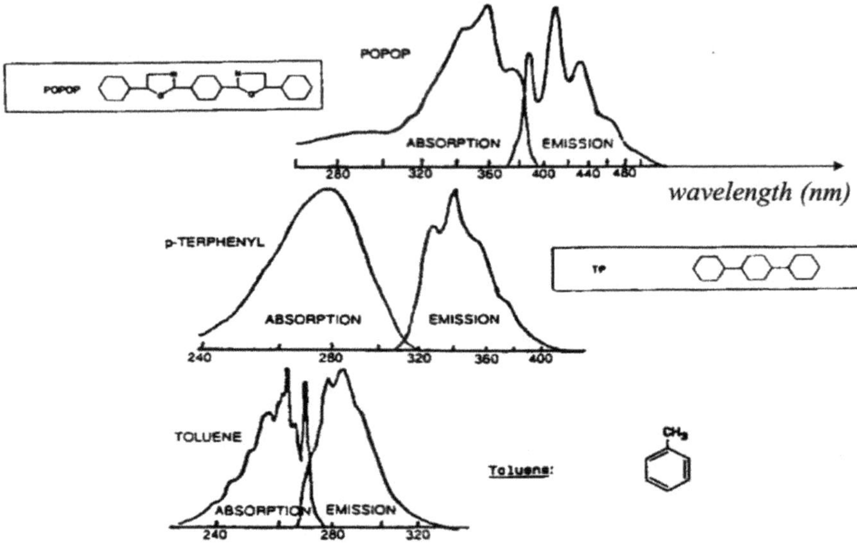

Figure 5.5. Binary scintillator, with two wavelength shifting compounds added to the main polymer (toluene). The resulting emission is in the visible range.

as shown in the example of Figure 5.5, achieving the best match with the spectral response of the photon detector.

Organic wavelength shifters belong to the family of aromatic compounds, composite chains of benzene rings; they are commonly identified with nicknames deriving from their chemical configuration: PT (*p*-terphenyl), PPO, PBO, BBQ, POPOP and similar. For an exaustive list of WLS compositions see foe example Ref. (Selove *et al.*, 1979). Added to plastic or liquid solvents, they constitute the basis of most plastic scintillators widely used in particle physics, medical diagnostics, astrophysics and other applications. Table 5.2 provides data on the basic WLS and on commercially available scintillator materials, named with an old terminology from Nuclear Enterprises; other manufacturers have equivalent products under different names. The exact composition of the materials is generally an industrial property.

Plastic scintillator can be fashioned to fit the experimental needs; Figure 5.6 shows a selection of shapes. Before use, the modules are wrapped with a dark coatings to prevent sensitivity to external light;

Table 5.2. Basic organic and commercial plastic scintillators (from different sources).

Organic scintillators	λ_{MAX} (nm)	τ_{DECAY} (ns)	Light output (% anthracene)
Anthracene	447	30	100
Stilbene	410	4.5	50
Plastic			
NE-102A	423	2.4	65
NE-104	408	1.8	68
Pilot F	425	2.1	64
Liquid			
NE-213	425	3.2	78
NE-323	425	4.0	60

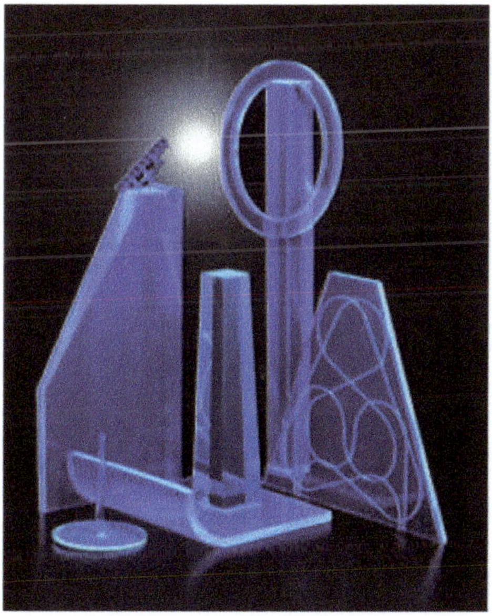

Figure 5.6. A collection of plastic scintillator slabs before wrapping and mounting.

photon sensors are connected to one or more sides, often with a transparent plastic light guide. Examples are shown schematically in Figure 5.7; Figure 5.8 is a disassembled set of a classic photomultiplier mounting.

Convenient to collect the signals from large arrays, such as those employed in particle calorimeters, is the use of WLS bars optically coupled

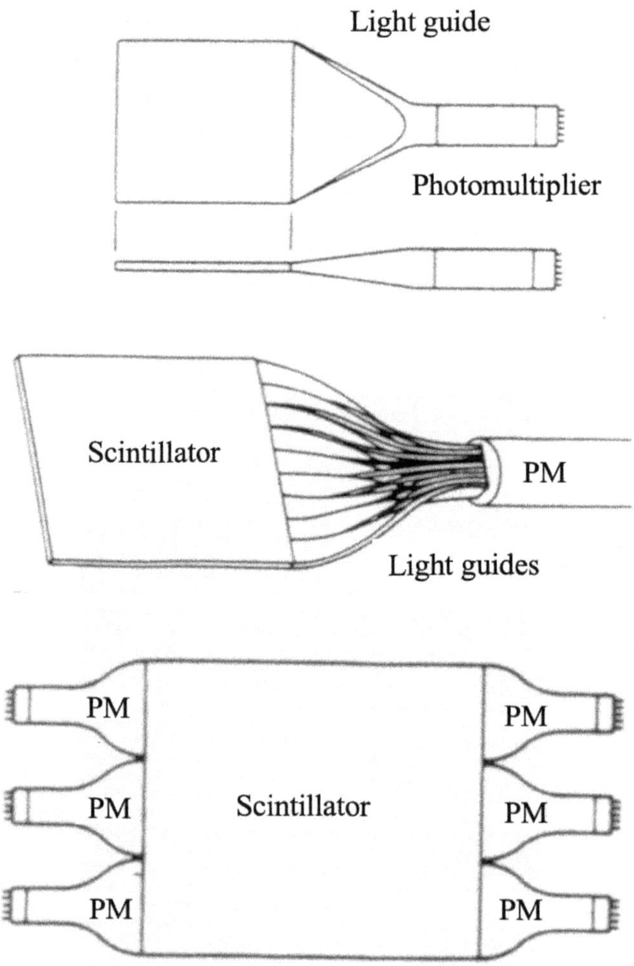

Figure 5.7. Examples of scintillator assemblies, with single or multiple light guides connection to the phototubes.

Figure 5.8. A classic set of scintillator, photomultiplier and base providing the voltage distribution.

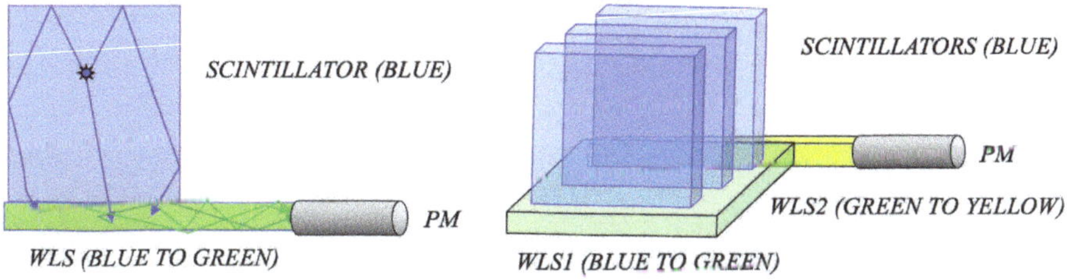

Figure 5.9. Wavelength shifter bars collect the light from a scintillator with a single (left) or double coupling (right).

to the scintillators, as shown schematically in Figure 5.9; fluorescence photons from the scintillator are absorbed in the WLS bar and re-emitted isotropically at longer wavelengths. The process can be repeated, covering large detection volumes with a single sensor. Figure 5.10 shows an assembled calorimeter module with multiple WLS bars readout.

In very large systems, the scintillation light can be collected and transported over long distances making use of large arrays of WLS optical fibres; an example is the ATLAS hadronic calorimeter, shown in Figure 5.11.

Normally employed for readout and signals transmission, plastic scintillating fibres have been also used as tracking detectors, packing a large number in bundles with position-sensitive readout at one end. Figure 5.12

Figure 5.10. A hadronic calorimeter prototype with scintillator stacks and multiple WLS readout bars (Picture CERN).

Figure 5.11. A module of the ATLAS hadronic calorimeter with WLS optical fibres readout (Picture CERN).

shows the cross section of an individual fibre, 60 μm in diameter, surrounded by six adjacent neighbours; Figure 5.13 is the image of a charged particle track crossing a detector with several thousand scintillating fibre bundles, recorded with a solid-state camera (D'Ambrosio *et al.*, 1993).

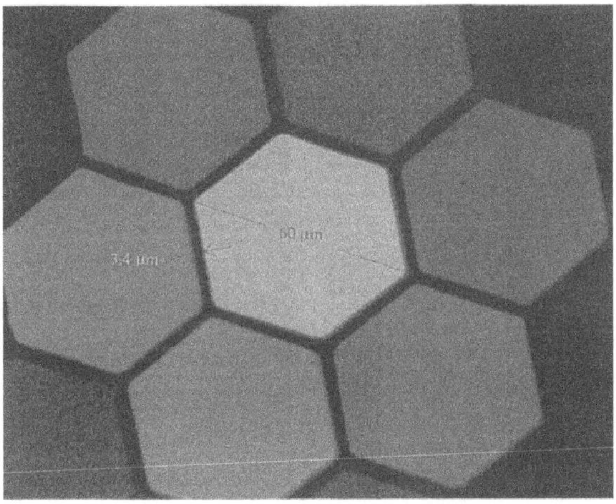

Figure 5.12. Closely packed scintillating fibres, 60 μm in diameter.

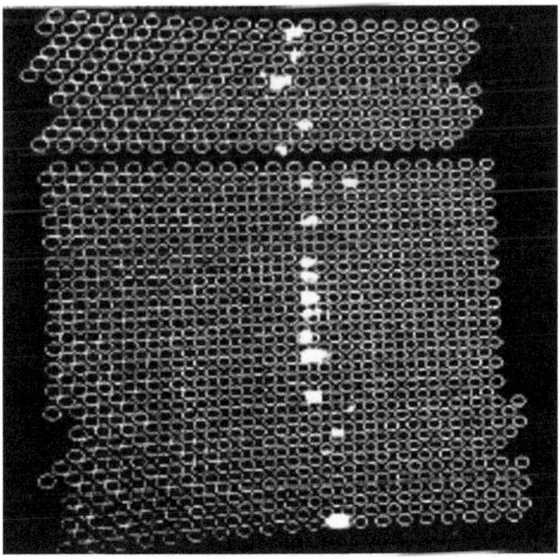

Figure 5.13. Charged particle track recorded with a solid-state camera (D'Ambrosio *et al.*, 1993).

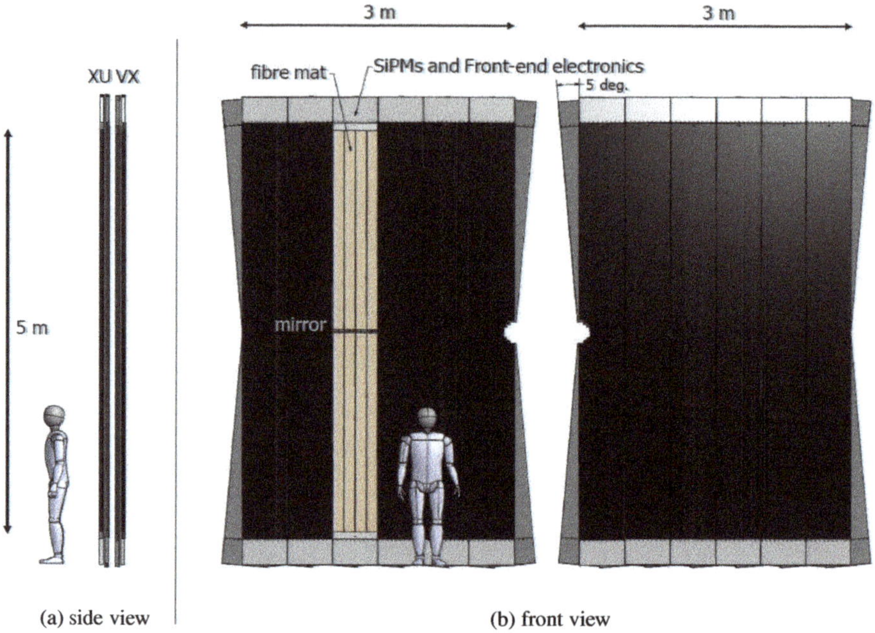

Figure 5.14. CERN's LHCb Scintillating Fibres upgrade (Joram *et al.*, 2015).

Large scintillating fibres arrays have been built for high-rate tracking in particle physics experiments. In the UA2 detector, operated from the mid 1980s at CERN's proton-antiproton collider, a cylindrical arrangement of some 60,000 fibres in 24 layers readout by CCDs were used for tracking and pre-shower measurement; a cylindrical arrangement of some 60,000 fibres in 24 layers was used for tracking and pre-shower measurement in the Central Fibre Tracker of the Fermilab D0 experiment. The LHCb SciFi (Scintillating Fibres tracker), part of a major upgrade of the LHCb detector to allow operation at the increased luminosity of CERN's LHC, includes around half a million scintillating fibres readout by silicon photomultipliers, Figure 5.14.

5.3 Photon Sensors: Vacuum Photodiodes, Photomultipliers and Microchannel Plates

The photoelectric effect was discovered in 1887 by Heinrich Herz exposing a metal electrode to ultraviolet radiation; since then, years of observations and progress permitted to develop a wide assortment of photosensitive materials. Particularly efficient in the visible range are alkali metals composites with low work function: monoalkali (Ag-O-Cs, Sb-Cs), bialkali (Sb-Rb-Cs, Sb-K-Cs) and multi-alkali (Na-K-Sb-Cs). Caesium iodide (Cs-I) and caesium telluride (Cs-Te) insensitive respectively above 200 and 300 nm, and are used for the so-called solar-blind devices. The compilation in Figure 5.15 provides, as a function of wavelength, the quantum efficiency of a large range of photocathode materials in the transmission mode (Hamamatsu, 2017).

The photocathode deposition processes have been perfected to achieve maximum sensitivity, and are technical property of the various manufacturers; they are entirely executed under high vacuum, as even trace amounts of oxygen or water can affect the result. An exception is caesium iodide, that can be briefly handled in air before being sealed in an inert atmosphere; this property has been exploited to manufacture UV-sensitive gaseous detectors for Cherenkov Ring imaging (Section 14.6).

The simplest photon sensor is the vacuum photodiode, shown in Figure 5.16 (top): a sealed glass envelope with two electrodes. The

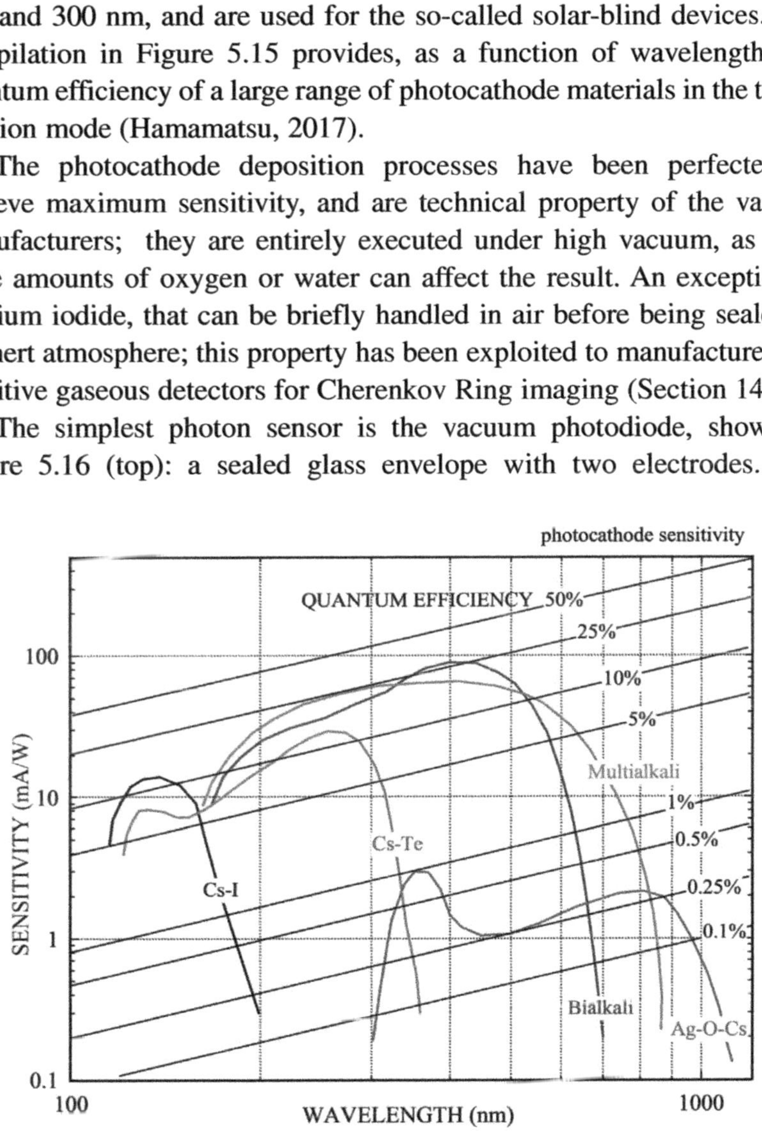

Figure 5.15. Quantum efficiency selected photocathodes (Hamamatsu, 2017).

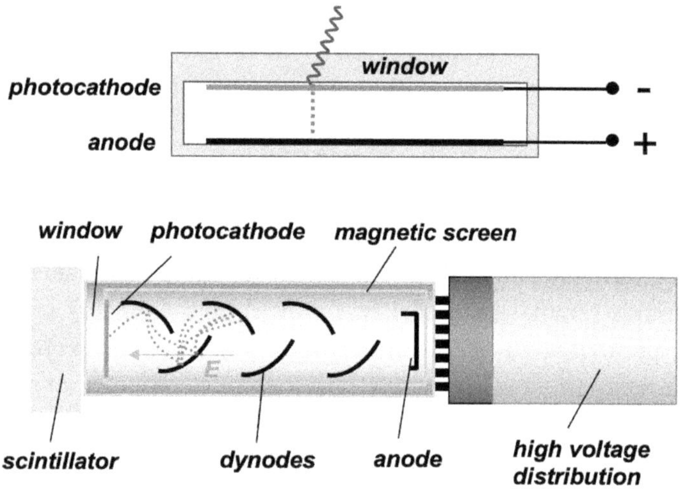

Figure 5.16. The vacuum photodiode (top) and the photomultiplier (bottom).

photosensitive material is deposited on the inner side of the window, usually coated with a thin transparent conducting layer (ITO, indium-tin oxide, or equivalent), providing the electrical contact to the generally insulating photocathode material. Having no gain, the photodiode is used to detect the current induced by high photon fluxes.

The photomultiplier, Figure 5.16 (bottom), is a more complex system providing high gains, permitting to detect low light levels, down to single photons. As shown schematically in the figure, the primary photoelectron is accelerated by the applied electric field towards an electrode, named dynode, to an energy sufficient to extract several secondary electrons; the process continues through a succession of dynodes resulting in a large signal on the final charge collecting anode.

While the photocathode material governs the spectral response of the tubes, the transparency of the tube entry window determines the region of sensitivity, particularly at short wavelengths; Figure 5.17 provides the transmittance of several materials used as windows. With a cutoff around 120 nanometres, magnesium fluoride is the most transparent material in the vacuum ultraviolet, only surpassed by lithium fluoride, difficult to use because of its hygroscopic nature, with its cutoff at ~100 nanometres.

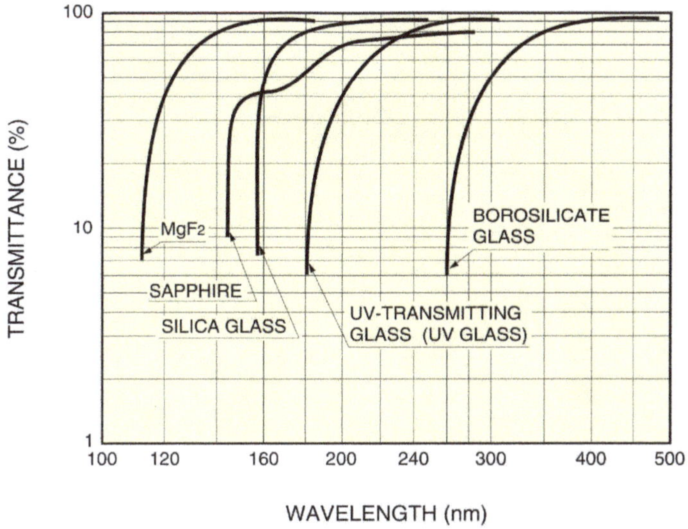

Figure 5.17. Transmittance of window materials as a function of wavelength (Hamamatsu, 2017).

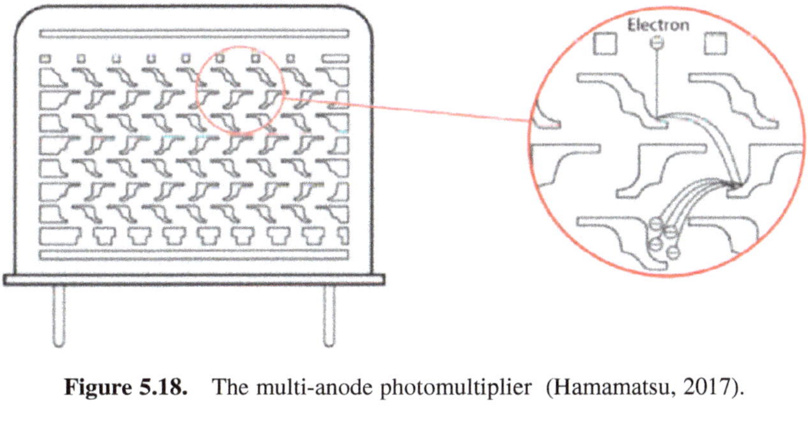

Figure 5.18. The multi-anode photomultiplier (Hamamatsu, 2017).

Differently from the single dynode chain used in standard photomultiplier, the multi-anode PM deploys a set of dynode grids in cascade, suitably arranged to amplify the photoelectrons over a large area, Figure 5.18. Generally of a square or rectangular shape, the MA-PM can have the charge collecting electrode segmented to perform localization of the primary charge.

Figure 5.19. A collection of phototubes in a wide range of sizes and performances (Picture Hamamatsu Photonics).

Commercial photomultipliers exist in vast range of sizes and performances. Figure 5.19 shows a collection of tubes, from Hamamatsu's current catalogue; other producers offer similar products. The largest model is the 50 cm diameter Hamamatsu R3600, used to instrument the Kamiokande water Cherenkov neutrino detector; over 12,000 tubes have been installed in the experiment, Figure 5.20 (Fukuda *et al.*, 2003.[1] The experiment was operational in 1987, when a supernova event was observed by optical telescope. The neutrinos arrived at earth on February 1987, and the Kamiokande detector observed 11 events.

Large arrays of photomultipliers are deployed in experiments designed to detect cosmic neutrinos in the TeV energy range from the Cherenkov radiation emitted in water or ice by the charged particles generated by the interactions of neutrinos in space, with the atmosphere or the medium. The Deep Underwater Muon and Neutrino Detector project (DUMAND), based 30 km offshore the island of Hawaii, started at the end of the seventies and planned to implement more than 200 optical modules arranged in eight strings 40 m apart (Roberts, 1992). Figure 5.21 shows one of the

[1] In 2002, the detector suffered a major accident that destroyed about half of the phototubes. Named Super-Kamiokande, it was entirely rebuilt and fully operational a few years later.

Figure 5.20. Inside the Super-Kamiokande neutrino detector.

Figure 5.21. The optical detector module of DUMAND.

detectors, with a set of large size photomultipliers encased in a pressure-tight spherical vessel.

Discontinued in the mid-nineties due to technical difficulties, DUMAND was a precursor for other projects such as ANTARES (Astronomy with a Neutrino Telescope and Abyss environmental Research project) off the coast of Toulon (France), the Deep-Underwater Cherenkov Detector in Lake Baikal (Russia), the European Kilometer Neutrino Telescope (KM3NeT) detector offshore Porto Palo (Italy) in the Mediterranean sea. On February 13, 2023, the KM3NeT research infrastructure recorded the highest energy muon observed so far, at 220 Peta electronvolts (1 PeV $=10^{15}$ eV) named KM3-230213A. Given its enormous energy and near-horizontal direction, the muon most probably originated from the interaction of a neutrino of even higher energy in the vicinity of the detector (Aiello *et al.*, 2025).

The IceCube experiment, Figure 5.22, comprises 5160 digital optical modules instrumenting a volume of over one km^3 of South Pole glacial ice

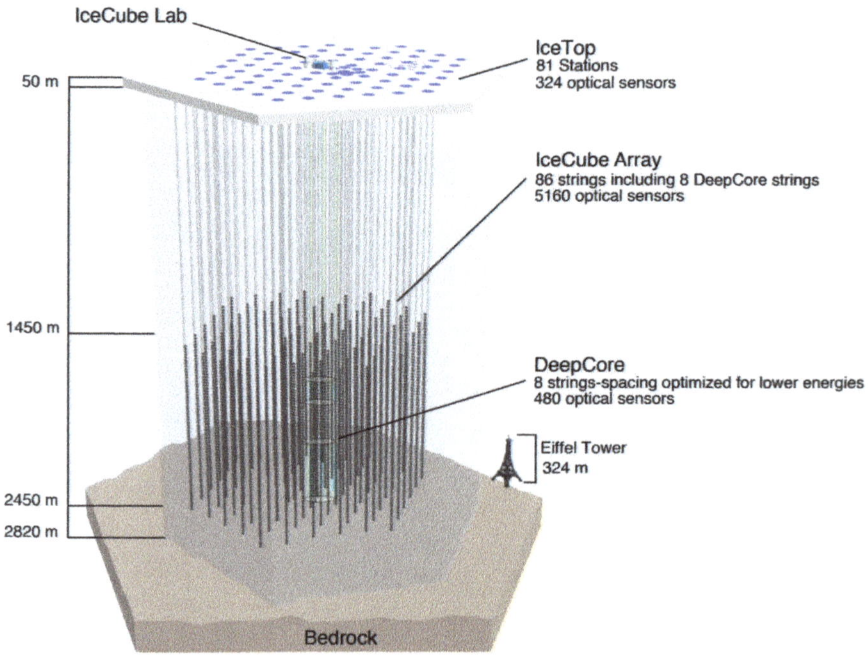

Figure 5.22. The IceCube detector (Yu *et al.*, 2023).

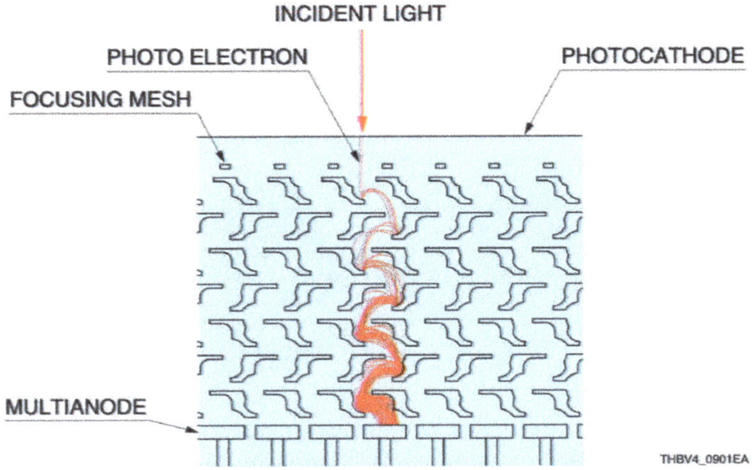

Figure 5.23. The MAPM ((Hamamatsu, 2017).

deep below the surface. Each module contains a photomultiplier tube and the associated electronics to detect and read out photons generated by Cherenkov photons produced by the relativistic charged particles propagating through the ice, produced by neutrino interactions.

The multianode photomultiplier (MAPM), Figure 5.23, is a vacuum device with a large number of narrow dynode channels; electrons emitted by the photocathode are transported by the electric field through a sequence of electrodes and undergo successive collisional multiplication until collected at the anode, single or segmented to perform localization. MAPM can have a compact square or rectangular shape, convenient to cover large detection areas. Figure 5.24 shows an example of a commercial 64-anodes tube.

The single channel multiplier, or channeltron, consists in a thin insulating tube coated on the inner walls with a secondary electron emitter; accelerated in vacuum by an electric field, electrons enter one side of the tube and multiply by successive collisions with the walls, Figure 5.25. To reduce the problems due to positive ions feedback, the tube can have a bent shape, and has been used as detector in mass spectrometers; in this case the charge generating the electron cascade is a positive ion.

The Microchannel plate (MCP) consists of a two-dimensional array of glass capillaries moulded together in the shape of a disk, Figure 5.26 (left). Each channel has an internal diameter ranging from 5 to 25 microns,

Figure 5.24. A 64-channels H8500 MAPM from Hamamatsu.

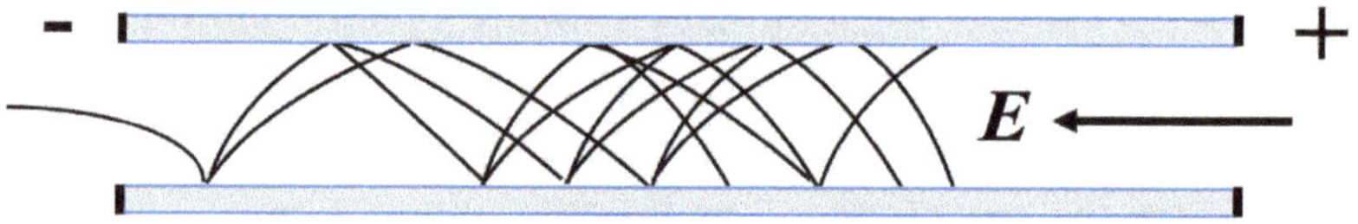

Figure 5.25. The single channel electron multiplier, or channeltron.

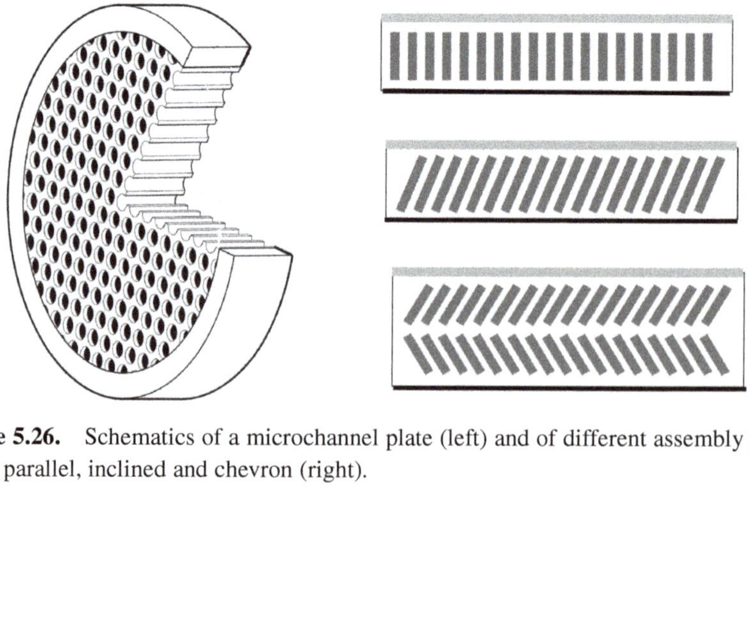

Figure 5.26. Schematics of a microchannel plate (left) and of different assembly geometries: parallel, inclined and chevron (right).

with the inner wall processed to have a suitable electrical resistance and good secondary emission properties. When a primary electron impinges on the inner wall of a channel, secondary electrons are emitted and accelerated by the electric field created by the voltage applied across the two ends of the MCP. The secondary electrons hit the channel wall to produce additional secondary electrons; the process is repeated many times and as a result a large number of electrons are released from the end of the channel. MCPs can be assembled in several geometries. For the simplest scheme with parallel fibres (top right) positive ions generated in the channels by electron-wall collisions can drift back and damage the photocathode; this is avoided using inclined fibres in the so-called chevron configuration (middle right). Two chevron plates in cascade with opposite inclination (bottom right) permit to obtain large gains.

5.4 Solid State Photon Detectors: Photodiodes, Avalanche Photodiodes, Hybrid Photomultipliers

A semiconductor is a crystalline material (silicon, germanium, gallium arsenide and others) having conductivity properties between metals and insulators. Even a superficial description of semiconductor properties is outside the scopes of this book, and the reader is referred to the vast number of dedicated books for more exhaustive explanations; some are suggested in the "Basic Bibliography". Only a short reminder is given here.

Silicon belongs to the IV group of elements, with 4 electrons on the valence shell. Excitation of an electron to the conduction band creates at the same time a hole, a fictitious entity representing the absence of an electron in the shell; electrons and holes move under the effect of an electric field and are the charge carriers in the semiconductor. In pure (intrinsic) silicon, the thermally created densities of electron and holes are equal, around 10^{10} cm^{-3} at room temperature. Doped *p*-type and *n*-type materials are obtained replacing typically 1 in 10^8 of the silicon atoms with atoms belonging to the III and V groups, having 3 and 5 valence electrons: they are named respectively acceptors and donors. In *p*-type semiconductors, adding trivalent acceptors (boron, gallium, indium), the attachment of electrons creates holes, which are the majority carriers;

in *n*-type, adding pentavalent donors (antimony, phosphor, arsenic), the majority carriers are electrons. The notations p^+, n^+, p^{++}, n^{++} denote increasing levels of doping as compared to the basic. In all materials there are carriers of opposite charge, generated by thermal excitation of silicon atoms; they are called minority carriers.

Doping of an intrinsic semiconductor is realized by diffusion with exposure to a gas of impurity atoms at high temperature, or bombarding the substrate with ions of the dopant (ion implantation). A *p-n* junction is then created with controlled doping of adjacent regions of the crystal with the two species; patterning and metallization of the contacts is done with high-precision photolithographic processing. For illustration purposes, Figure 5.27 provides an example of the manufacturing process of a silicon diode.

The silicon photodiode, Figure 5.28, is a *p-n* junction semiconductor with a depletion region *i* created by the difference of potential between anode and cathode. Electrons and holes liberated by photons or ionizing radiation in the depletion region are collected on the respective electrodes. With its band gap of ~1.2 eV, silicon converts photons in the few eV energy range (UV to infrared) into charge with efficiencies nearing 60–80%; the *p* implant is kept as thin as possible to reduce absorption. With no gain and a noise nearing 1000 electrons, the photodiode is used mostly for integral light flux measurements.

In the more complex structure of the avalanche photodiode (APD), Figure 5.29, the charge can be largely increased by an internal electron multiplication process. With an applied very high reverse bias on the *p-n*

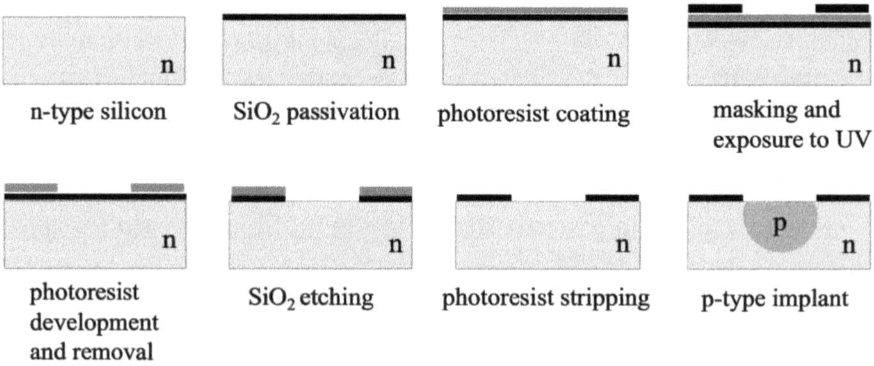

Figure 5.27. Manufacturing process of a p-n junction diode.

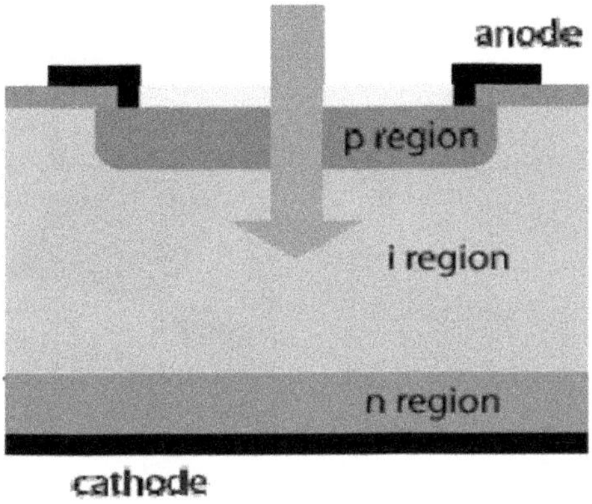

Figure 5.28. Schematics of a silicon photodiode; electron-hole charges released by ionizing tracks in the depleted region *i* are collected on the electrodes.

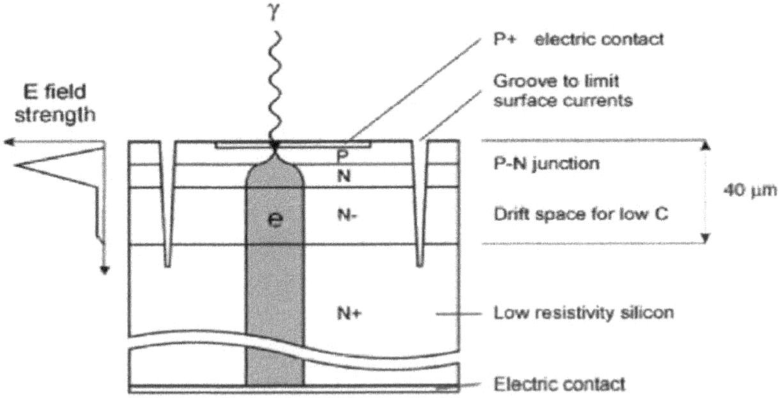

Figure 5.29. Structure of the avalanche photodiode. Photoelectrons created in the p-type layer are multiplied in an avalanche process in the high field of the *p-n* junction and drifted to the collecting electrode.

junction, the structure generates a collisional avalanche multiplication process for electrons resulting in proportional charge gains around 100 before breakdown. APDs are compact, efficient and radiation hard; they have been chosen to readout the scintillation light in the CMS electromagnetic calorimeter. A double APD module produced by Hamamatsu is shown in Figure 5.30 (Ingram *et al.*, 2001).

Increasing the operating voltage of the APD above proportionality, an operating regime analogous to the Geiger discharge in gaseous

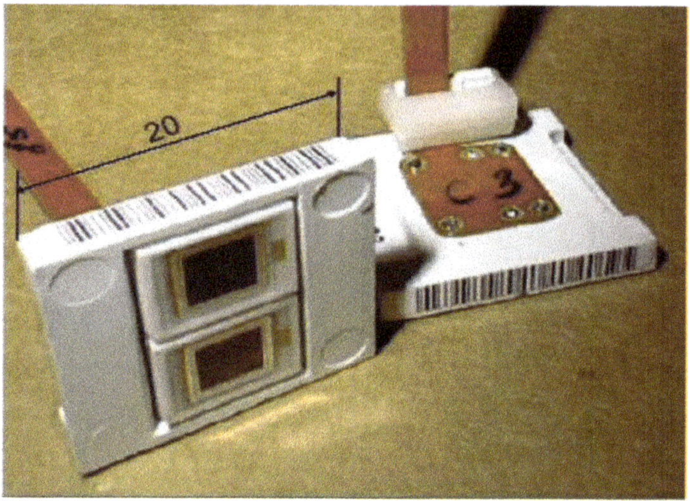

Figure 5.30. A double APD module for the CMS electromagnetic calorimeter (Ingram *et al.*, 2001).

counters provides conveniently large pulses; however, the diode's large capacitance results in very long recovery times and possibly permanent damages. Proposed in the early 2000 and further developed by a group at the Moscow Physics Institute (Buzhan *et al.*, 2003), the Silicon Photomultiplier (SiPM), also named Multi-Pixel Photon Counter (MPPC) or Single-photon Avalanche Diode (SPAD), is a semiconductor device consisting of hundreds to several thousand independent APDs built on a common silicon substrate individually operated in a limited Geiger mode; resistors connect the diodes to the substrate limiting the current, Figure 5.31. While the individual diodes provide saturated signals, the sum of pulses is proportional to the incident photons count, see Figure 5.32. The SiPM can be conveniently mounted on the tip of optical fibres collecting the fluorescence photons of scintillators, see Section 5.2. For a recent overview of the technology, see for example, (Piemonte and Gola, 2019).

SiPM are commercially available in a vast range of sizes and packaging, as shown for example in Figure 5.33.Their high sensitivity, low cost and good aspect ratio make the SiPM an ideal device to instrument medical diagnostics systems, PET in particular (Bisogni *et al.*, 2019).

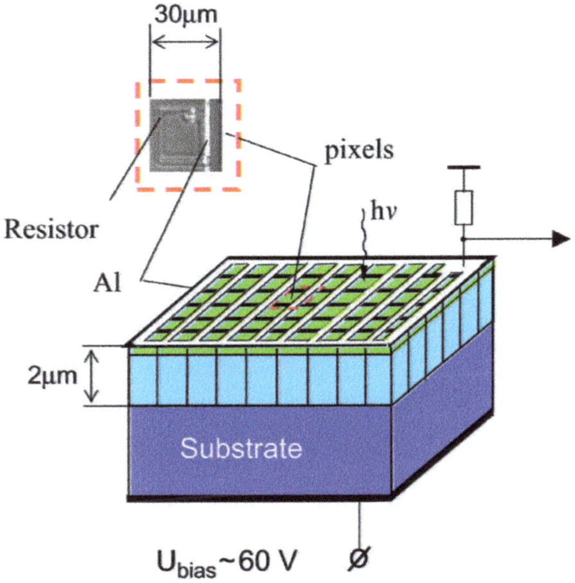

Figure 5.31. Schematics of a silicon photomultiplier (Otte *et al.*, 2005).

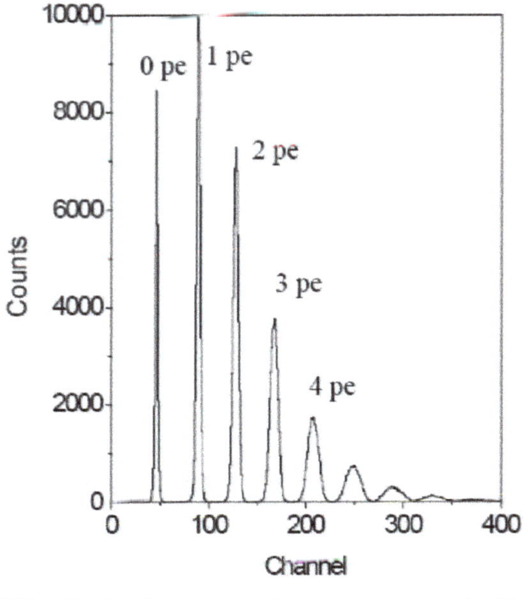

Figure 5.32. Single photons counting spectra recorded with a SiPM.

Figure 5.33. Examples of silicon photodiodes (from Hamatsu Photonics).

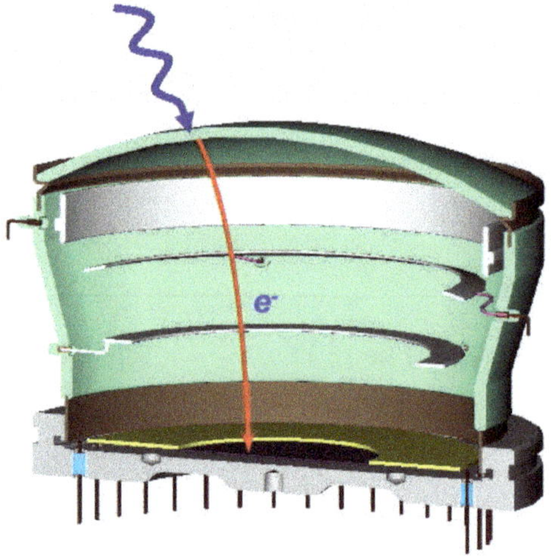

Figure 5.34. Structure of a Hybrid Photon Detector (HPD).

The Hybrid Photon Detector (HPD), coupling in the same vacuum casing a photocathode with a solid-state sensor, permits to attain the large gains needed to detect and localize single photoelectrons (Figure 5.34). The electrons emitted from the cathode exposed to light are accelerated by an arrangement of electrodes at suitable potentials and focused on a sensor. Operating the device at a difference of potential of 15–20 kV imparts to the electrons an energy high enough to provide large signals in the silicon sensor (D'Ambrosio *et al.*, 1994), that can be a simple diode or a more complex position-sensitive matrix of silicon pixels (Section 6.2). Large

Figure 5.35. A large HPD with silicon pixels readout developed for CERN's LHCb experiment (Picture CERN).

Figure 5.36. Photon detector for the LHCb RICH upgrade (Picture CERN).

area hybrid photomultipliers have been developed by the LHCb collaboration at CERN to perform particle identification detecting the Cherenkov ring pattern generated in a radiator by fast charged particles (Gys, 1999, Albrecht *et al.*, 2000), see Section 14.6; Figure 5.35 is one of the large HPDs arrays used in the experiment.

Starting in 2019, to cope with the increased events rate provided by the higher collider luminosity, the LHCb HPD Ring Imaging Cherenkov Counter (RICH) has been replaced with a photon imaging system based on the use of multi-anode photomultipliers (MAPMT) with 8 × 8 pixels. Figure 5.36 shows one of the photodetectors planes, with 2-inch tubes in the low occupancy region, and 1-inch tubes in the higher rate central region (Fiorini, 2020).

Chapter 6

Solid State Radiation Detectors

6.1 Silicon Microstrips

Silicon detectors have been in use for years in particle physics. Developed in the eighties, the planar fabrication technology permits to manufacture multi-electrodes devices, providing detection and localization of charged particles at the 5–10 μm level (Kemmer, 1980). For a detailed description of the manufacturing processes and operating characteristics of the devices see for example the 1999 book by Gerhard Lutz (1999).

As shown schematically in Figure 6.1, a microstrip detector is built using photolithographic tools on an *n*-type thin silicon plate, few hundred microns thick; *p*+-type strips are formed by diffusion or ion implantation, capped by aluminium contacts. With the application of suitable potentials, fully depleted diodes are formed between *p*+-type strips and the n^+ backplane; charges induced by the motion of electrons and holes released by ionizing encounters are sensed on the metal strips, connected to amplifying electronics.

Figure 6.2 shows an early silicon microstrip detectors built and operated in 1980 for the NA11 experiment at CERN, 400 μm thick, with 100 strips at 200 μm pitch and fast current sensitive preamplifiers connected to each strip. A charge-sharing algorithm on the recorded signals provided position accuracies below 10 μm for minimum ionizing particles (Heijne *et al.*, 1980).

With the emerging availability of large-scale integrated circuits for signals readout and recording, silicon microstrips detectors underwent a

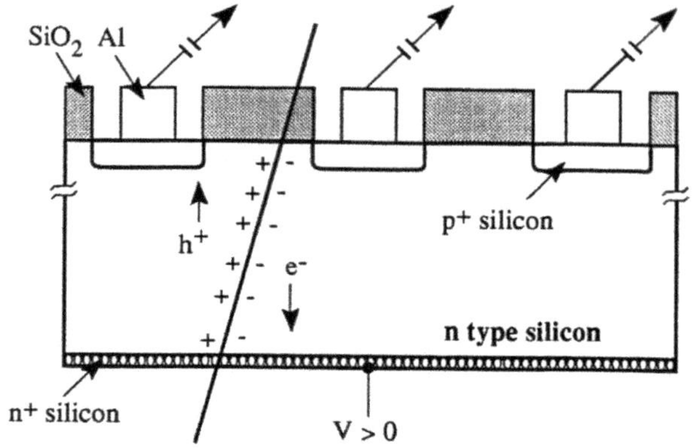

Figure 6.1. Schematic drawing of a silicon microstrip detector.

Figure 6.2. The NA11 silicon strip detector at CERN ((Heijne *et al.*, 1980).

swift progress, targeted at the needs of particle physics experimentation. The historical overviews by Michal Turala and Manfred Krammer (Turala, 2005, Krammer, 2009), and Frank Hartmann's book (see Hartmann, 2009) describe these early advancements. The design and manufacturing of dedicated high-density readout chips reversed the sizes ratio between active detector and electronics, as seen in Figure 6.3 showing the readout end of a module of the DELPHI vertex detector at CERN's Large Electron Positron collider (LEP) (Chochula *et al.*, 1998); a comparison with

Figure 6.3. Close-up of DELPHI silicon strips detector, using 128-channels large-scale integrated readout chips (Picture CERN, 1996).

Figure 6.2 illustrates well the progress in electronics. Figure 6.4 is the assembled microvertex detector of the ALEPH experiment at CERN (Decamp *et al.*, 1990), operated from 1989 to 2000, when LEP was discontinued to permit the construction of CERN's Large Hadron Collider (LHC).

Figure 6.4. The ALEPH silicon microstrips vertex detector (Picture CERN).

Further evolution of the dedicated large scale integration led to the development of the APV25 chip, broadly used for silicon microstrips and gaseous micro-pattern detectors readout; the circuit includes a 128-channels low-noise amplifiers, an analogue pipeline and a readout multiplexer (French *et al.*, 2001). As seen in Figure 6.5, showing a module of the CMS silicon strips detector, the ratio of active to total surface of the detector has improved considerably as compared to earlier implementations.

Designed to operate at CERN's LHC, the Compact Muon Solenoid (CMS) vertex detector, shown schematically in Figure 6.6, has an inner three-layers pixel detector, described in the next section, and ten layers of silicon microstrips of increasing diameter; two end-cap systems complete the tracker (CMS_Collaboration, 2008). The detector, comprising 15,000 silicon strips modules and a sensitive are of more than 200 m^2 (Figure 6.7) was installed in 2007 and operated until 2016, when an upgrade was needed to cope with the increased luminosity. Similar systems instrument the other LHC experiments at CERN, ALICE, ATLAS and LHCb.

Figure 6.5. Module of the CMS silicon microstrips detector, with the APV25 readout chips and two interconnected 6-inch detectors (Picture CERN, 2000).

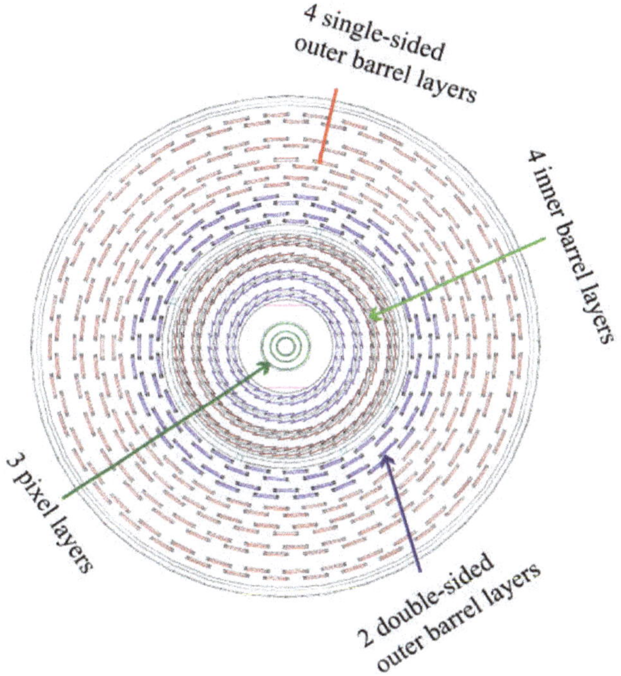

Figure 6.6. Cross section of the CMS vertex detector (Sguazzoni, 2008).

6.2 Silicon Pixel Detectors: CCD, MAPS, Hybrid

During the last decade of the 20th century, the increasing particles multiplicity created by interactions at the new generations of accelerators necessitated the elaboration of novel detector designs to instrument the core of particle physics experiments. Named vertex detectors, the new devices comprise a large number of silicon diodes with a pitch of a few tens of microns, individually readout by dedicated electronic circuits. Three major architectures have been developed, schematically depicted in Figure 6.8: Charge Coupled Devices (CCD), Monolithic Active Pixel Sensor (MAPS) and Hybrid Pixels; each type has several variants of design. Recent articles by Eric Heijne (Heijne, 2021) and Chris Damerell (Damerell, 2023) review the history and evolution of pixel devices, as well as their applications in particle physics and medical imaging.

Figure 6.7. One half of the CMS silicon strips inner barrel detector assembled in San Piero a Grado (Pisa) before shipping to CERN (2006).

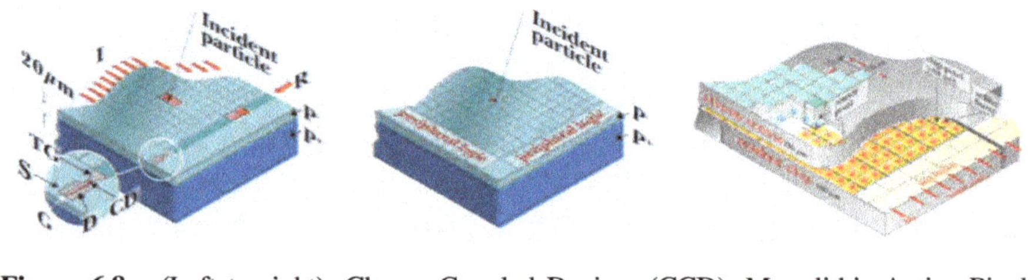

Figure 6.8. (Left to right): Charge Coupled Devices (CCD), Monolithic Active Pixel Sensor (MAPS) and Hybrid Pixel Detector (Damerell, 2021).

Developed in the late 1960s by Willard Boyle and George E. Smith and capable of recording visible light images, CCD comprise arrays of linked photosensitive capacitors built on silicon with integrated circuits technology that, under control of external applied potentials, can transfer their charge along rows and columns of adjacent cells to a common output, Figure 6.9. CCDs of various designs are widely used for applications

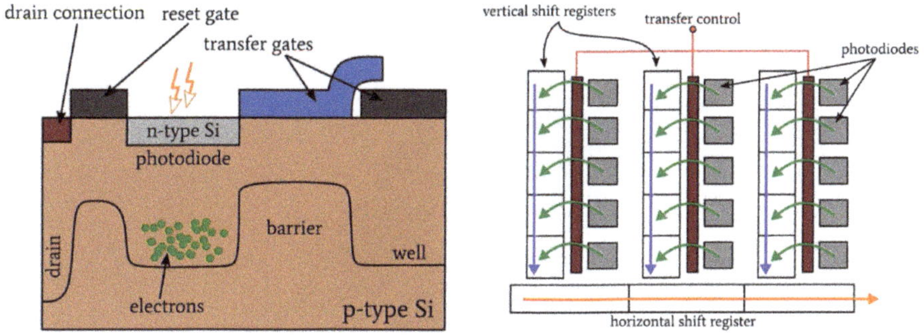

Figure 6.9. Left: layout of a CCD pixel structure; right: the charge transfer sequence (Keim, 2020).

in photography, television and consumer's electronic imaging; the development of multi-megapixel sensors has revolutionised the field of optical imaging in astronomy and astrophysics. Boyle and Smith were awarded the Nobel Prize for Physics in 2009 for their invention.

Although not specifically designed to detect ionizing radiation due to their thin depletion layer, CCDs have been used in particle physics increasing the sensitive layer thickness to 10–20 μm; this results in the release by minimum ionizing particles of over 10^3 electron-hole pairs, a charge well above the intrinsic pixel noise of around 100 electrons. A CCD-based tracker has the unique properties of ambiguity-free recording of multiple tracks, with a few micron two-dimensional position determination; due to the relatively long time needed to shift the information, their use in experiments is most advantageous to detect events with moderate rate and high multiplicity interaction yields. For a detailed description of early developments and applications see the review of Chris Damerell (Damerell, 1998).

A detector comprising 96 large CCDs, ~80 × 16 mm^2 each, with a pixel size of 20 × 20 μm^2 has been operated from 1996 in the SLD vertex detector at the Stanford Linear Accelerator Centre (SLAC). Figure 6.10 shows one half of the ~300 Mpixel VXD-3 vertex detector assembly, and Figure 6.11 is a schematic cross section in the X–Y plane, with recorded multiple tracks of an event superimposed.

The detector functioned until the end of the SLD operation in 2002 and contributed significantly to the high-precision quantum chromodynamics studies in the Z_0 mass region.

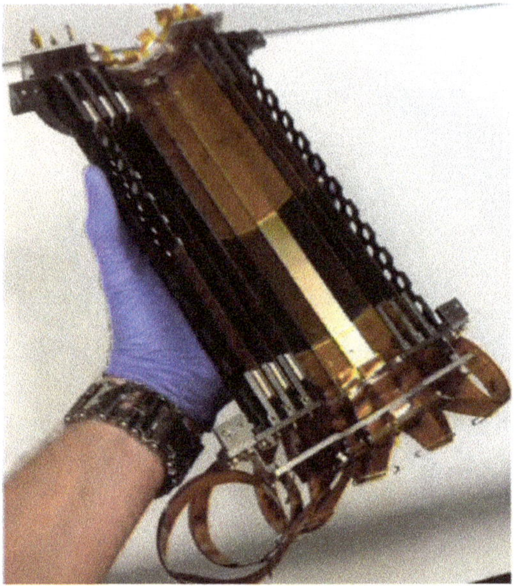

Figure 6.10. One half of the SLD-VXD3 CCD-based vertex detector. Two identical modules are assembled around the vacuum tube of the collider (Damerell, 2021).

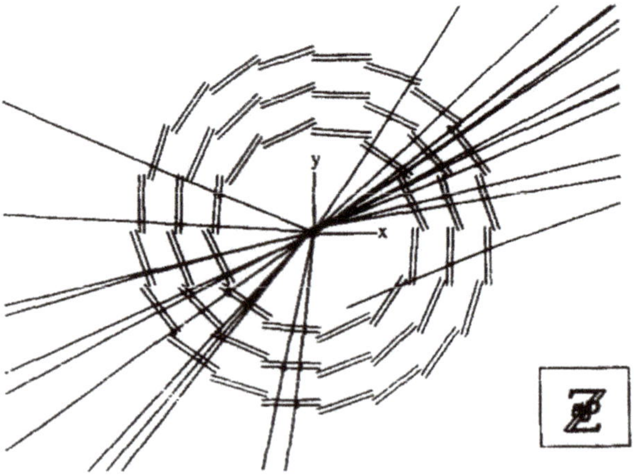

Figure 6.11. View of the SLD-VXD3 vertex detector with a recorded event superimposed (Abe *et al.*, 1997).

Despite their high granularity and low cost, the use of CCD-based pixel detectors is limited by the relatively long events readout time, tens of milliseconds. Years of dedicated research and the availability of large-scale application-specific circuits led to the introduction of hybrid devices, associating a matrix of silicon pixel detector to front-end readout circuits; the two elements are interconnected using a bump-bonding technology, shown schematically in Figure 6.12. This approach permits to optimize independently the manufacturing technologies and performances of sensors and readout electronics.

Hybrid detectors are designed to handle the high radiation environment and provide fast events readout needed to instrument the core of high-energy detectors. The first implementation of the CMS experiment inner tracker had three coaxial sensors barrels with radius of 4.3, 7.2 and 11 cm, about 50 cm long, surrounding the accelerator pipe and deploying 66 million pixels with 100 × 150 μm granularity; Figure 6.13 shows the detector components before installation.

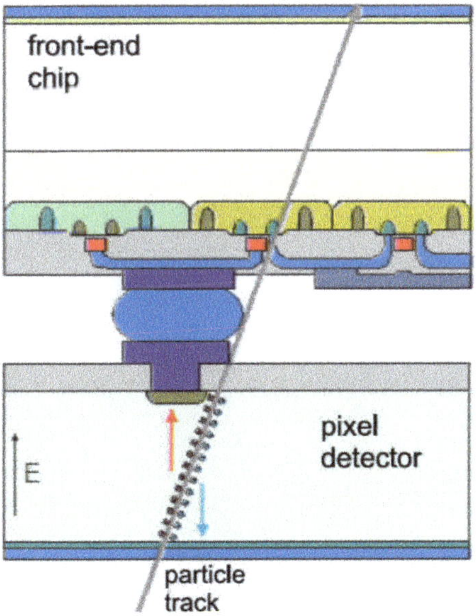

Figure 6.12. Structure of a hybrid pixel detector (Marco-Hernández, 2020).

Figure 6.13. CMS inner tracker detector prior to installation (Picture CERN).

To cope with the increasing LHC luminosity and replace defective and radiation-damaged modules, the detector underwent an upgrade during the 2016–2017 machine shutdown; a new upgrade is foreseen in 2025 (Adam *et al.*, 2021).

In Monolithic Active Pixels Sensors (MAPS), specifically developed for high-rate particle physics applications, each pixel has its own charge-integrating detection circuit and signals are sequentially read out from each on-chip switching circuitry; Figure 6.14 is an example of a pixel

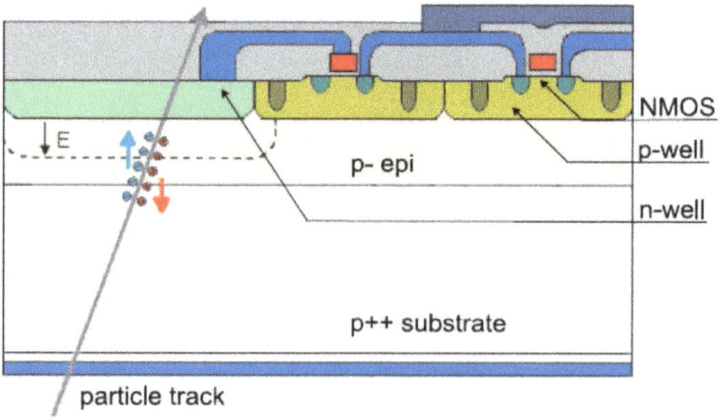

Figure 6.14. Example of a MAPS pixel realization (Riedler, 2015).

Figure 6.15. The STAR MAPs-based pixel detector at RHIC (Contin, 2017).

implementation. The simplest readout scheme is the "rolling shutter", in which peripheral logic along the chip edge addresses rows in turn, and analogue signals are transmitted by column lines to the bottom of the imaging area. Figure 6.15 shows the STAR vertex detector at Brookhaven's Relativistic Heavy Ions Collider (RHIC), a system featuring 356 MPixels 20×20 μm in size, operated from 2014 (Contin, 2017).

During the 2019–2021 LHC shutdown, a MAPS-based inner detector has been installed as major upgrade of the ALICE experiment at CERN to replace the previous setup based on hybrid modules (ALICE_Collaboration, 2024).

6.3 Miscellaneous Solid-State Detectors

Developed in 1984 by Emilio Gatti and Pavel Rehak, and functionally equivalent to its gaseous counterpart, the semiconductor drift chamber (or Silicon Drift Device, SDD) utilizes the transport properties of electrons in a thin silicon wafer in which a drifting field has been superimposed to the depletion field to drag the ionization electrons to a collecting anode (Gatti and Rehak, 1984). Figure 6.16 shows schematically a semiconductor drift chamber structure with sideway depletion field.

SDD are used in particle physics to instrument the inner core of several accelerator experiments, as the ALICE vertex detector at CERN and

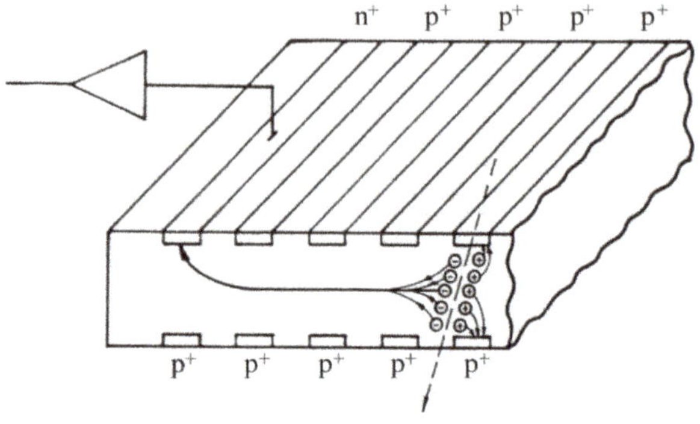

Figure 6.16. Schematics of the silicon drift chamber (Gatti and Rehak, 1984).

the STAR silicon vertex tracker at BNL. As shown in the picture of Figure 6.17, individual SDD plates are assembled in "ladders", mounted radially around the collider (Bellwied *et al.*, 2003). The transverse track position is obtained from a measurement of the drift time, while rows of segmented anodes provide the longitudinal coordinate.

Figure 6.17. The Silicon Vertex Tracker of the STAR experiment at BNL.

Due to their very low capacitance, SDD have a low intrinsic noise and are therefore suited for applications requiring very good energy resolution as in X-ray spectroscopy. Optimized for astrophysics applications, Figure 6.18 show a SDD module with cylindrical field geometry and an integrated amplifier on the central collecting anode (Curceanu *et al.*, 2019). A range of high-resolution Silicon Drift Detectors is commercially available (*PNSensor GmbH*, 2018).

Originally proposed by S. Parker and collaborators (Boldyrev *et al.*, 2016), the 3D solid state sensor consists in a 3D arrangement of holes penetrating the detector's bulk at a few tens of μm pitch, alternatively filled with *n*-type and p-type silicon, and forming a distributed array of diodes (Figure 6.19). The short charge collection distance for the

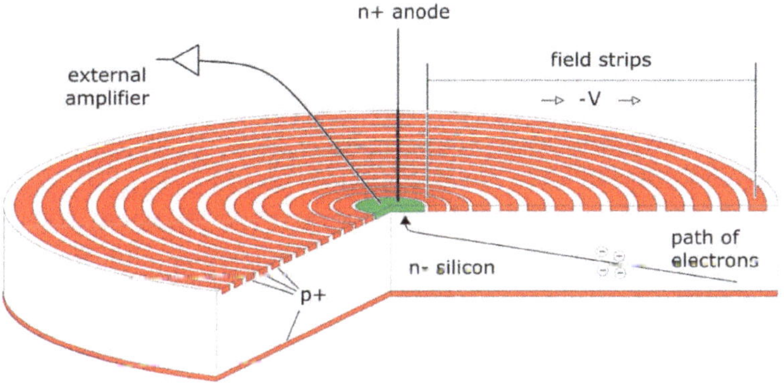

Figure 6.18. Design of a SDD module optimized for applications in astrophysics (Curceanu *et al.*, 2019).

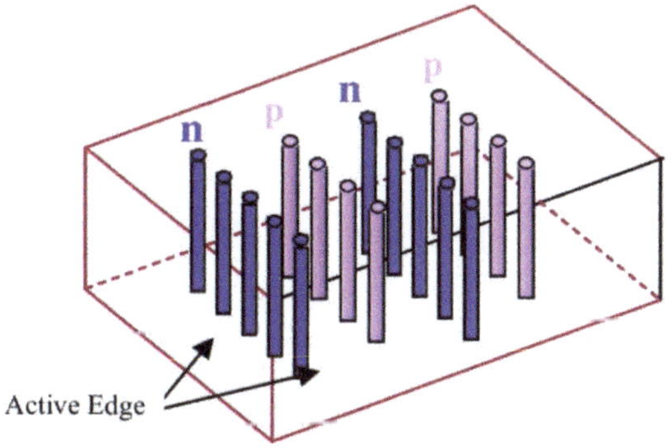

Figure 6.19. Structure of the 3D silicon sensor (Terzo *et al.*, 2021).

ionization charges results in sub-ns time resolutions and better radiation tolerance, thanks to the higher applicable electric field that reduces the trapping probability of radiation-induced defects. For detailed studies of manufacturing technologies and of major performances in detecting and localizing high-rate charged particles see (Parker *et al.*, 1997, Da Via *et al.*, 2012) and the book by Cinzia Da Via, Gian Franco Dalla Betta and Sherwood Parker (Da Via *et al.*, 2019). The manufacturing process, better described in the quoted references, is only schematically described here: starting with the bulk material, column electrodes are dug from one side with reactive ions etching and filled by surface reaction with n^+ and p^+ doped silane. Further photolithography processing forms the contact electrodes. A system of 3D pixel sensors has been operated from 2012 in the ATLAS inner tracker, and an upgrade with improved radiation tolerance to face the challenge of the Higher Luminosity LHC is in preparation. Figure 6.20 shows the pixel structure of the ATLAS upgrade (Da Via, 2012). With their high radiation tolerance and sub-μm position accuracy, 3D sensors find applications for imaging in medical diagnostics, neutron imaging and other fields.

Developed from the late 1990 in view of applications for medical imaging, the Medipx–Timepix hybrid chips family and consist in a square matrix of active pixels with low noise amplifiers, discriminators

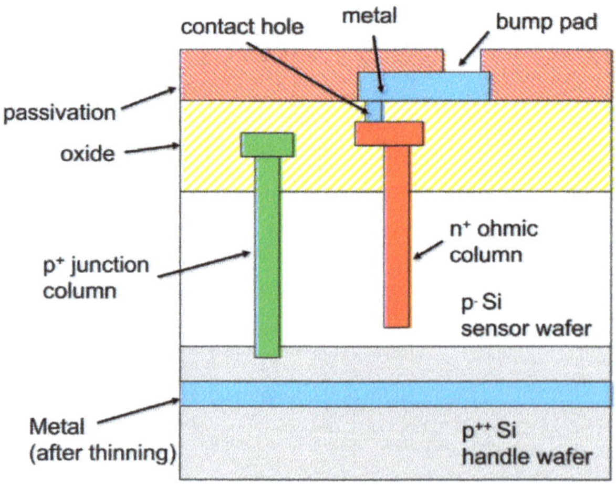

Figure 6.20. Structure of a 3D silicon pixel (Dalla Betta and Povoli, 2022).

and readout shift registers and has single photon counting capability (Terzo *et al.*, 2021).

From the 64 × 64 pixels of the early designs, the chip design evolved to 256 × 256 pixels in Medipix2 and Medipix3 with 55 μm pixel size; bump-bonded to a silicon sensor, the chip permits to record 2D images of ionizing radiation (Figure 6.21). With a modified readout structure, the Timepix (Figure 6.22) permits to record the arrival time of the ionization charges, and is therefore apt to provide time-resolved 3D images (Llopart *et al.*, 2002).

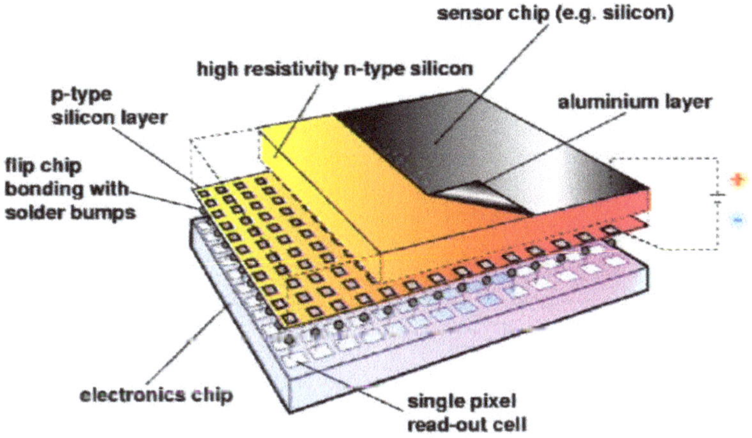

Figure 6.21. The Medipix–Timepix chip, coupling a single-pixel readout electronics structure to a solid-state sensor.

Figure 6.22. The Timepix3 sensor board, with 512 × 512 square pixels 55 μm on the side; the sensitive area is about 30 × 30 mm^2 (Picture CERN).

The evolution and performances of the Medipix–Timepix families, as well as several applications in spectroscopic X-ray, gamma-ray and particle imaging are described by (Llopart *et al.*, 2007).

The Medipix and Timepix chips have been used for radiation recording coupled to a gaseous amplifying structure, both the Micromegas and the Gas Electron multiplier (GEM); named GRIDPIX and GEMpix, thanks to the charge multiplication factor of the gaseous devices, the range of sensitivity of the devices is extended to low ionization losses.

Chapter 7

Proportional and Geiger Counters

The wire counter was developed in the early 1900 by Ernest Rutherford and Hans Geiger to study natural radioactive elements; as seen in Figure 7.1, it consists in a metal tube with a thin wire (the anode) stretched between two insulating plugs. A gas-tight vessel with a wide bore stop-cock permits to inject alpha particles emitted by a radium source mounted on a sliding rod. On application of a difference of potential between anode and cathode, the ionization charge is amplified by a charge multiplication process; the counts detected by the needle of an electrometer were found to correspond to the estimated activity of the source.[1] The detector operated at low pressures, both to extend the range of the α particles in the gas and to permit operation with the voltage sources available at the time.

Increasing the voltage further, the counter provides very large, saturated pulses in the so-called Geiger operating mode, with photon-mediated discharges propagating along the whole tube, and is capable to detect ionization yields down to single electrons. A high value serial resistor damps the discharge, avoiding irreversible damages, but has as consequence a limited rate capability.

Developed in the late nineteen twenties by Geiger and Walter Müller, Geiger–Müller detectors have and still are commercialized as affordable radiation monitors in a wide range of applications; their sensitivity to

[1] The term "proportional counter" was introduced several years later, when the availability of charge-recording electronics permitted to determine the correlation between the released ionization with the detected charge.

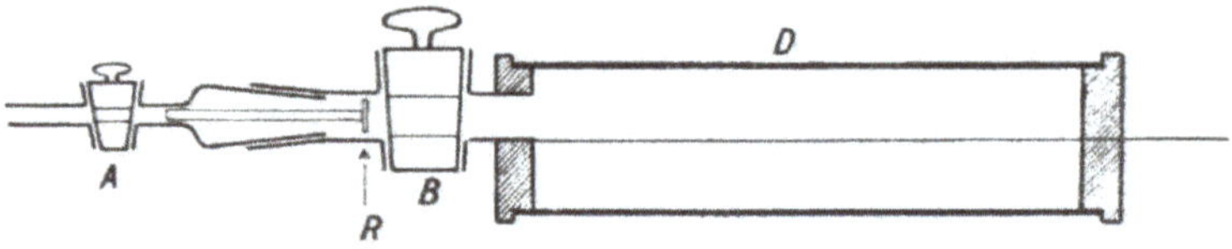

Figure 7.1. The proportional gaseous counter (Rutherford and Geiger, 1908).

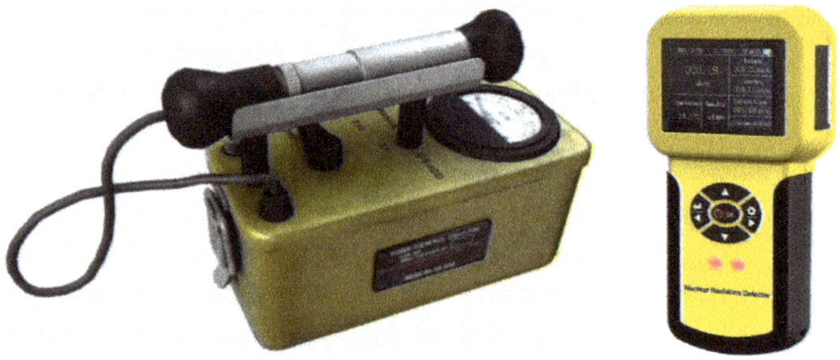

Figure 7.2. Examples of commercial Geiger-Müller counters. Today gaseous radiation monitors are being replaced by solid-state devices, easier and cheaper to manufacture.

radiation is only limited by the absorption in the front window. Figure 7.2 shows examples of commercially available counters.

With a variety of gas fillings and pressures, and instrumented with electronic recording circuits, single wire proportional counters are a major tool for the detection and identification of radiation: X-rays, charged particles and neutrons.

Figure 7.3 shows the dependence of detected charge from the voltage applied between anode and cathode for two sources of radiation having different ionization yields q_1 and q_2. After a steep rise, due to recombination losses, the collected charge reaches a plateau in the so-called ionization chamber regime; above a threshold voltage charge multiplication sets in, increasing exponentially the detected charge until reaching saturation. The gain, or multiplication factor of the counter, is defined as ratio of the detected charge to the primary ionization yield, Q/q; it can be measured or calculated from numerous expressions (see for example the book by Serge A. Korff in 1955).

Due to the structure of the electric field, having values below the critical value for charge multiplication in most of the sensitive volume except

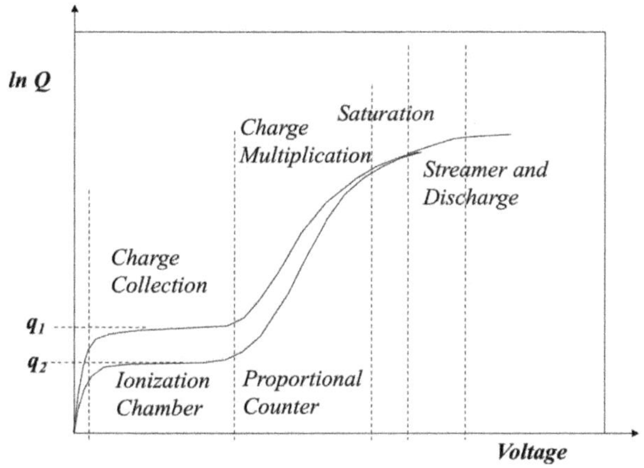

Figure 7.3. Dependence of recorded charge from operating voltage for two sources of ionization.

at a short distance from the anode wire, the detected charge is proportional to the ionization yield.

The energy resolution of a proportional counter is a convolution of three major dispersive factors: the electronics noise, whose contribution decreases with gain, the primary ionization statistics and the fluctuations in the avalanche process, increasing with gain. As shown by the example of Figure 7.4, part of a study aiming at finding the conditions for best resolution, the optimum is obtained for moderate value of gain (Järvinen and Sipilä, 1982).

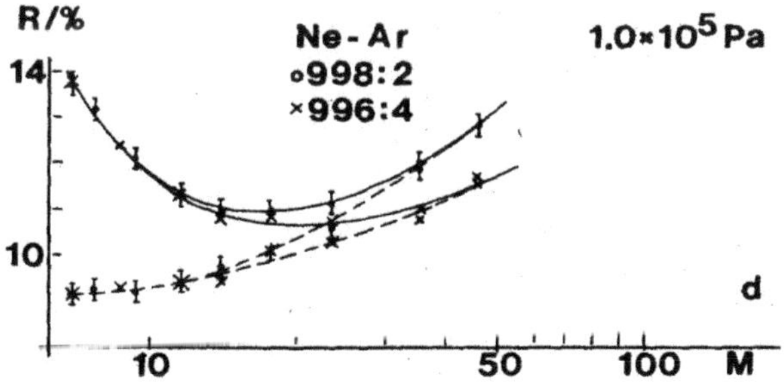

Figure 7.4. Resolution for 5.9 keV X-rays as a function of gain in a counter with neon-argon mixtures (Järvinen and Sipilä, 1982).

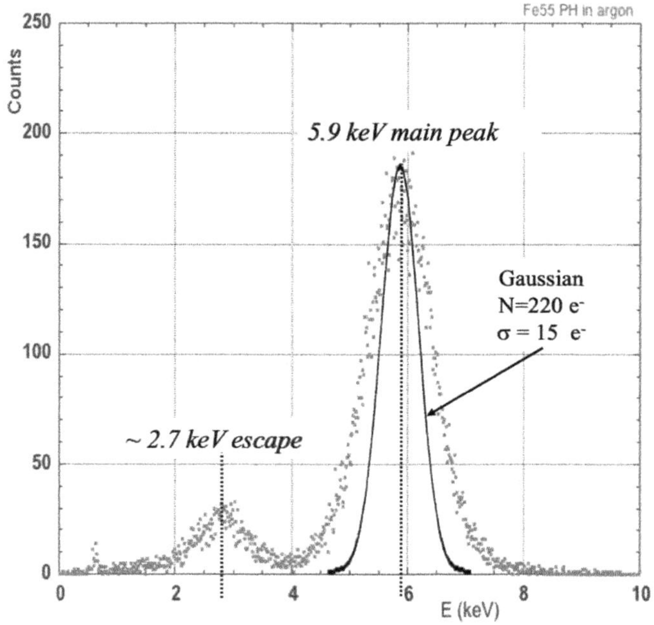

Figure 7.5. Energy loss spectrum of a ^{55}Fe source in argon. The inset shows the statistical dispersion for the main 5.9 keV line.

Shown in Figure 7.5 is the energy loss spectrum measured with an argon-methane proportional counter exposed to an ^{55}Fe radioactive source; the main peak is the sum of K- and L-shell absorptions, while the peak around 2.7 keV (the escape peak) corresponds to conversions where a 3.2 keV fluorescence photon did not reconvert within the sensitive volume. The inset shows the estimated contribution of a Gaussian energy loss of 220 electrons; the intrinsic resolution is degraded by the avalanche multiplication statistics.

Developed in the seventies, scintillating proportional counters exploit the copious emission by some gases of photons at electric fields below the onset of charge multiplication, thus avoiding the dispersive contribution of the avalanche statistics. The best results have been obtained with a spherical shape anode, coupled to a photomultiplier to record the gas scintillation; an example is given in Figure 7.6 for the detection of 5.9 keV X-rays (Policarpo *et al.*, 1972). Having its K-shell energy around 34 keV, the escape peak is absent in xenon-filled counters.

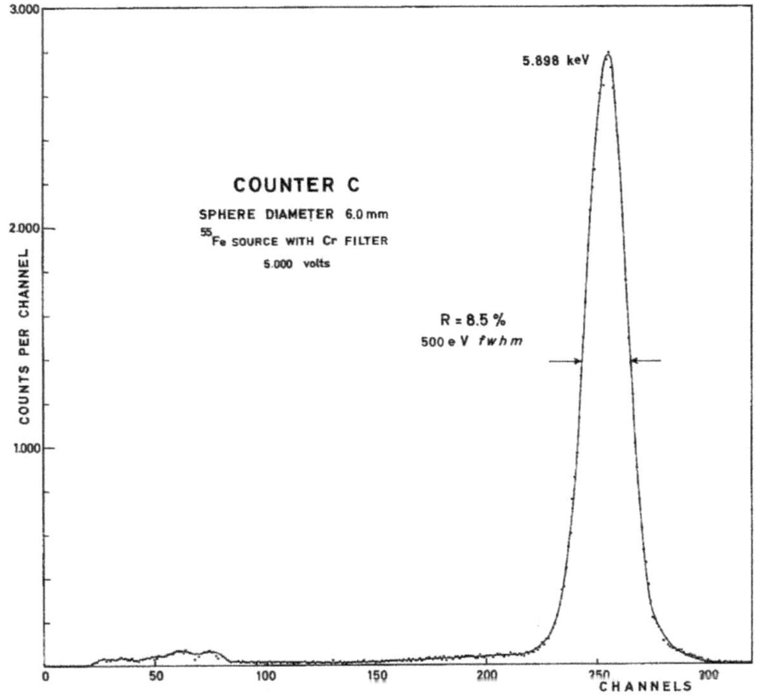

Figure 7.6. 5.9 keV X-ray line detected with a xenon scintillating proportional counter. The measured resolution (8.5% fwhm) is close to the statistical limit of 6.5%.

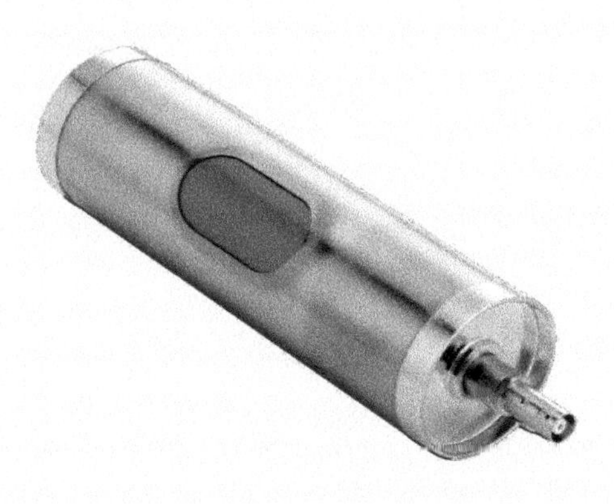

Figure 7.7. A xenon-filled counter with beryllium window (Type 45432 LND Inc, Oceanside New York, USA).

Figure 7.8. Thin window proportional counters (Canon Electron Tubes & Devices, Tokyo, Japan).

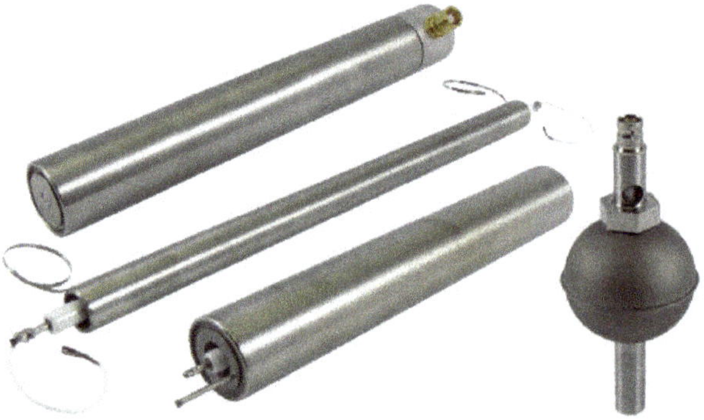

Figure 7.9. ^{3}He-filled counters for detection of neutrons (Global Nucleonics, Centronic LCC Huston, Texas, USA).

Single-wire proportional counters are commercially available in a wide range of choices and gas fillings for detection of X-rays, gamma rays, charged particles. The counter in Figure 7.7 is a xenon-filled model with thin beryllium window, optimized to detect soft X-rays above ~2 keV; Figure 7.8 is a range of counters produced by Canon, and Figure 7.9 is a line of ^{3}He-filled counters for detection of neutrons.

Chapter 8
Multiwire Proportional Chambers

The Multiwire Proportional Chamber (MWPC), developed in 1968 by Georges Charpak at CERN, revolutionized the field of radiation detectors. Comprising a set of thin parallel anode wires a few mm apart between two electrodes, the device operates as a set of individual proportional counters, with high counting rate capability and fast electronic readout (Charpak *et al.*, 1968). Equipotential and field lines generated by the application of symmetric negative potentials on the cathodes with the wires at ground potential are shown in Figure 8.1. Ionization electrons created in the sensitive volume drift to the high field region close to the anodes and are amplified by an avalanche multiplication process. The motion of charges created in the avalanche induces a negative signal on the anode and a positive charge distribution on surrounding electrodes.

In recognition of the impact of the new technology in basic and applied research, Georges Charpak received the 1992 Nobel Prize in Physics. Figure 8.2 shows Charpak's technician, Roger Bouclier, with the first MWPC prototype; the chamber had a ~10 cm^2 active area, with 30 wires at 3 mm pitch, connected by cables to the readout electronics.[1]

[1]The development the MWPC was presented by Charpak at the International Symposium on Nuclear Electronics in Versailles, September 10, 1968. Following the directives at the time for French citizens, the talk was given in French; in view of the interest raised, Charpak repeated later the presentation in English. I had attended the Symposium to present my work on magnetostrictive spark chambers; favorably noted by Charpak, a few months later, I was invited to join his research group at CERN.

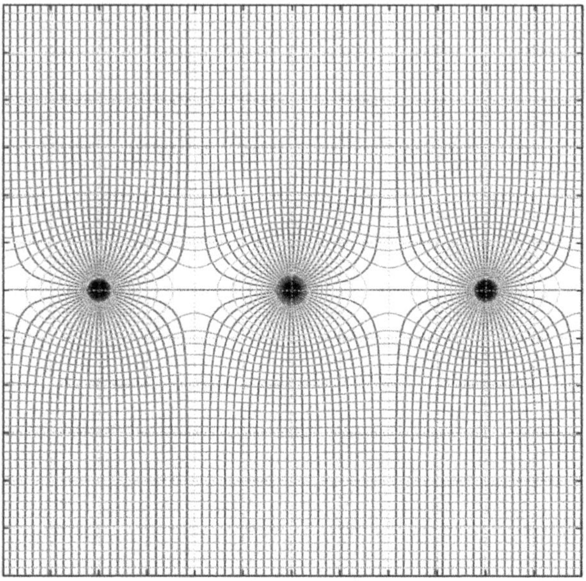

Figure 8.1. Mechanical structure, field lines and equipotential in the MWPC.

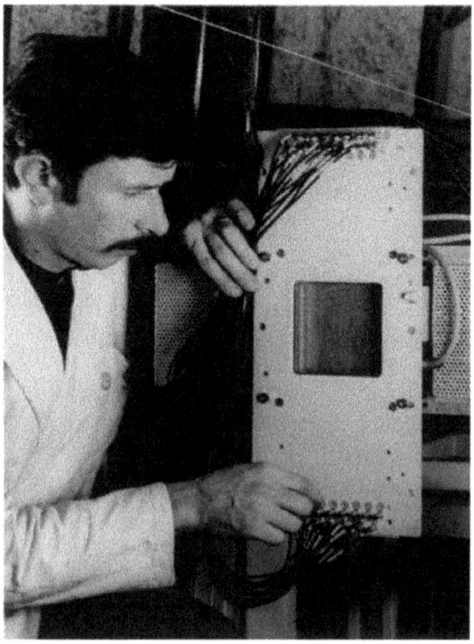

Figure 8.2. A MWPC prototype having 30 wires at 3 mm spacing (Picture CERN, 1967).

The position accuracy, in its basic design determined by the wire spacing, can be improved to better than a hundred microns with the development of a method to record the signals induced by the primary avalanche on segmented cathode strips, shown schematically in Figure 8.3 (Charpak *et al.*, 1968). With its unambiguous two-dimensional hit reconstruction, the cathode readout system is well suited to record multiple track events. The sub-mm space localization property has been exploited in medical imaging applications; in the detection of neutral events (X-rays and neutrons), the anode's signal can be used for triggering and energy-resolved recordings.

Owing to their exceptional operating characteristics, permitting to record events at rates orders of magnitude higher than the prevailing detectors at the time, MWPCs were swiftly developed and used in a new generation of experimental setups. The picture in Figure 8.4 shows a large size MWPC built for CERN's experiment PS17, a study of the rare decays of the neutral kaon to a muon pair (Schilly *et al.*, 1970). Figure 8.5 is a prototype for the Split Field Magnet (SFM) detector at CERN's Intersecting Storage Rings (ISR); it is also the first known picture of Charpak with a MWPC. The author of this book, a new member of the group, stands in the middle.

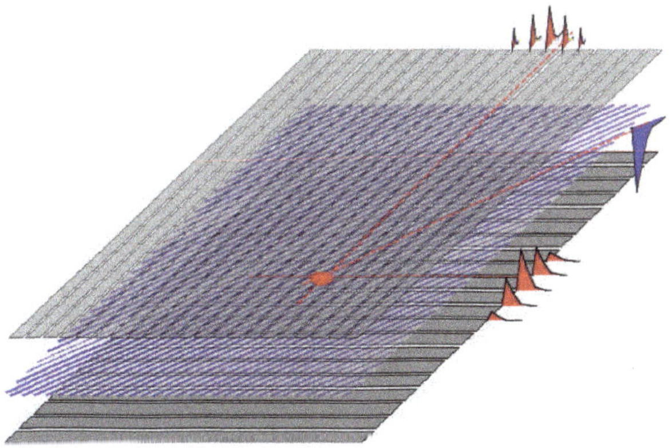

Figure 8.3. Two-dimensional MWPC readout. Charges created by the avalanche process generate a negative pulse on the anode wire and a wide profile of positive-induced charge on cathode strips, exploited for localization (Charpak *et al.*, 1968).

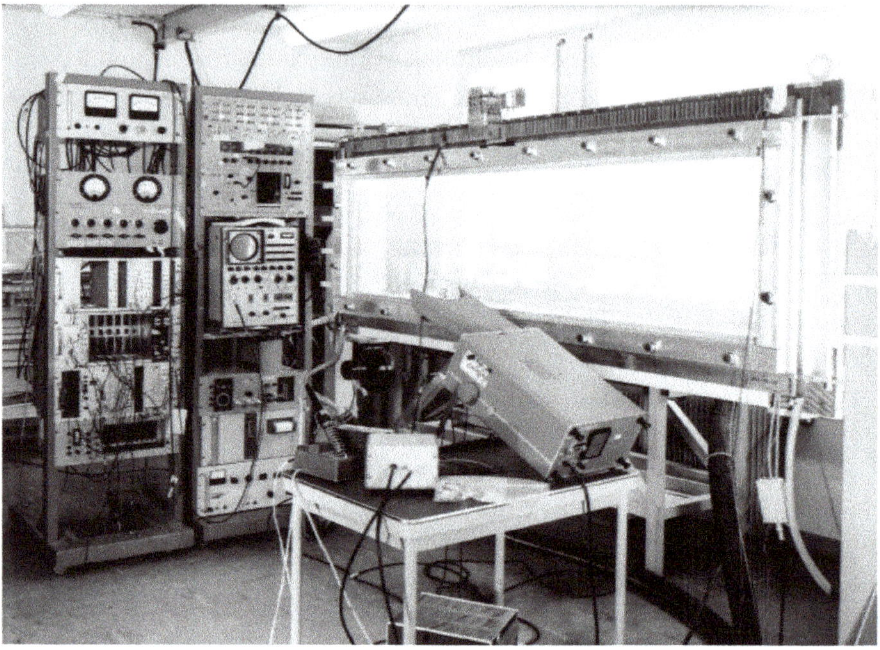

Figure 8.4. Large size MWPC for CERN's experiment PS17 (Picture CERN, 1970).

Figure 8.5. Georges Charpak (left) with a large size MWPC. Fresh member of the group, Fabio Sauli, stands in the middle (Picture CERN, 1971).

Figure 8.6. A lightweight MWPC prototype for the SFM detector (Picture CERN, 1972).

An innovative manufacturing technology, making use of light plastic foam supporting plates instead of heavy frames, was developed to optimize the sensitive detector area, paramount for use inside a magnet; Figure 8.6 shows a prototype of the SFM light MWPC, and Figure 8.7 is a view of the completed system including 20 modules, 100×200 cm^2 in size, encompassing about 50,000 readout wires (Bouclier *et al.*, 1974).

The experiment operated for more than a decade, until the decommissioning of the collider in 1984. While contributing to the study of the physics of proton–proton collisions at high energies, one of the design goals, the detection of quarks, was not achieved: These basic constituents of matter are known now not to exist in a free status.

Figure 8.7. The Split Field Magnet Detector at CERN's Intersecting Storage Rings (Picture CERN, 1974).

Chapter 9

Drift Chambers

9.1 Planar Drift Chambers

As observed in the early publications on MWPCs, the time taken by the ionization electrons to reach the anodes can provide information on the distance of the track from the wire; knowledge of the time of the encounter, or time zero, generally provided by external triggering devices, is required. A measurement of the drift time permits to achieve sub-mm localization accuracies with wire spacings of centimetres, reducing the number of electronics channels as compared to MWPCs, at the cost of a reduced time resolution.

The Multiwire Drift Chamber (MWDC) developed in 1971 by Albert H. Walenta and collaborators is shown in Figure 9.1. A supplementary field wire at negative potential, placed between anodes, improves the charge collection; the intrinsic right-left ambiguity is resolved using pairs of close anode wires, or with reconstruction algorithms making use of several detectors mounted at angles. A set of large MWDC was employed by the CERN-Heidelberg group in a magnetic spectrometer for the measurement of the branching ratio $K^+ \rightarrow e^+\nu/K^+ \rightarrow \mu^+\nu$ (Walenta *et al.*, 1971) and achieved 350 microns space resolution with two centimetres wire spacing.

In the original MWDC design, due to the non-uniform electric field, the space-time correlation needed to reconstruct the track position is non-linear, causing dispersive effects. An improved detector design, named High-accuracy Drift Chamber (HDC), was developed by the CERN group (Charpak *et al.*, 1973), Figure 9.2. A set of cathode wires at graded

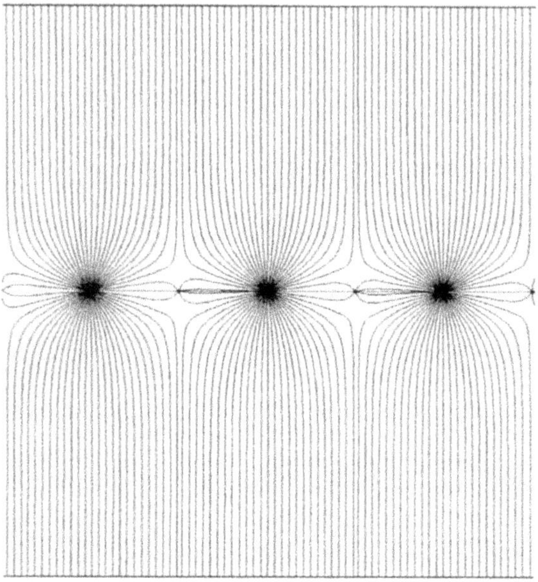

Figure 9.1. Electric field lines in the Multiwire Drift Chamber, with a field wire between anodes (Walenta *et al.*, 1971).

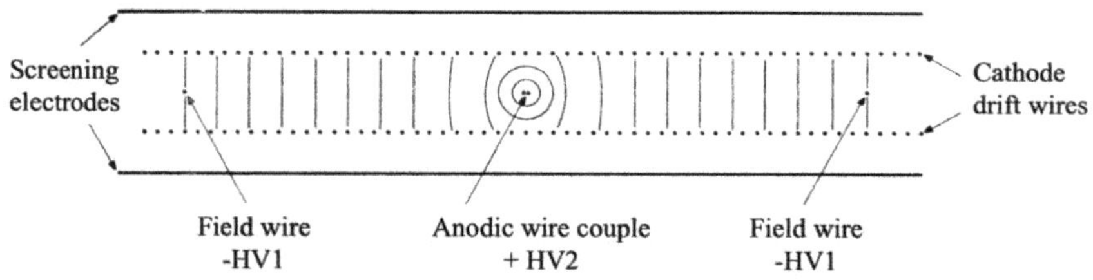

Figure 9.2. High-accuracy Drift Chamber (Charpak *et al.*, 1973).

potentials encompass a thin gap with uniform field, drifting ionization electrons to the anode; this geometry ensures a linear space-time correlation, and provides a position accuracy between 50 and 80 microns over the 50 mm cell size. Figure 9.3 shows a large size HDC built and operated by the OMEGA spectrometer group at CERN.

Making the most of their excellent space resolution, a set of medium size HDC was used successfully in experiments at CERN to study the channelling effects of fast charged particles in crystals (Fich *et al.*, 1975).

Figure 9.3. Large size high-accuracy drift chamber for the OMEGA spectrometer (Picture CERN, 1976).

A system of drift chamber constituted the main tracking device of the UA1 experiment at CERN's proton-antiproton collider (Baussgessner *et al.*, 1987). As shown in Figures 9.4 and 9.5, the detector was a complex of 40 modules in a cylindrical assembly, inserted in a magnet, with additional devices to perform particle identification. The experiment operated between 1980 and 1990 and led to the discovery of the $W^{\pm}$ and Z^0 bosons, mediators of the weak interactions; it was also one of the first experiments to use on-line tools for recording and displaying events.[1] The picture in Figure 9.6 is an example of event showing tracks recorded with the drift chambers; colours correspond to different particles nature and energy deduced from the particle identifiers.

[1] As far as I am aware, this was the first particle physics experiment to use commercial Macintosh computers for data handling.

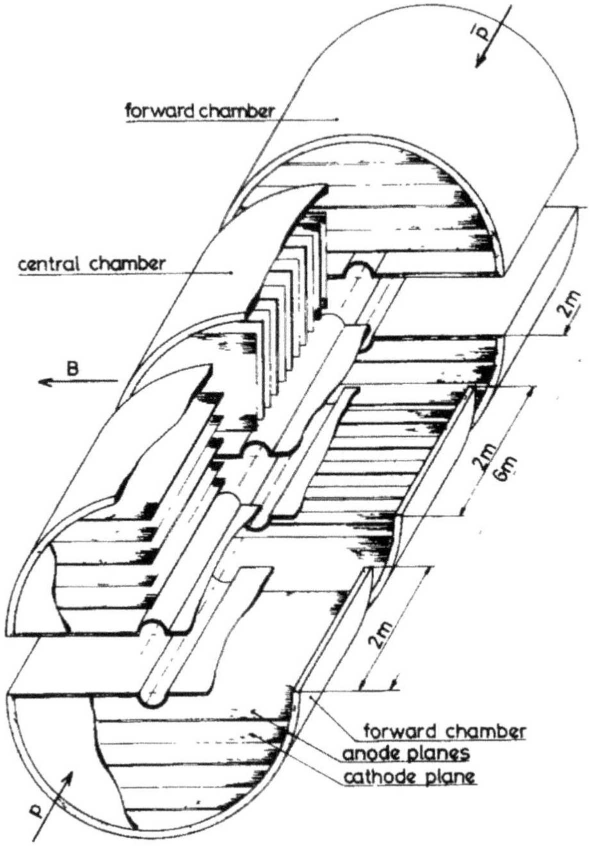

Figure 9.4. The UA1 drift chamber central detector (Baussgessner *et al.*, 1987).

The 1984 Nobel Prize for Physics was awarded to Carlo Rubbia and Simon Van der Meer for their contribution to the discovery.

The drift cell structure shown in Figure 9.7 with anode wires at 6 cm and additional field wires at potentials to provide a strong and uniform electric field for large cell sizes, was developed aiming at the construction of very large detectors, with anode wires at 6 cm and additional field wires at suitable potentials to provide a strong and uniform electric field for large cell sizes (Marel *et al.*, 1977). The wire structure is supported by light honeycomb panels, serving also as screening ground and gas containment.

Figure 9.8 shows the detector of the neutrino experiment WA1 at CERN, operated from 1978; the set-up had alternate layers of magnetized

Figure 9.5. The UA1 central drift chambers detector (Picture CERN).

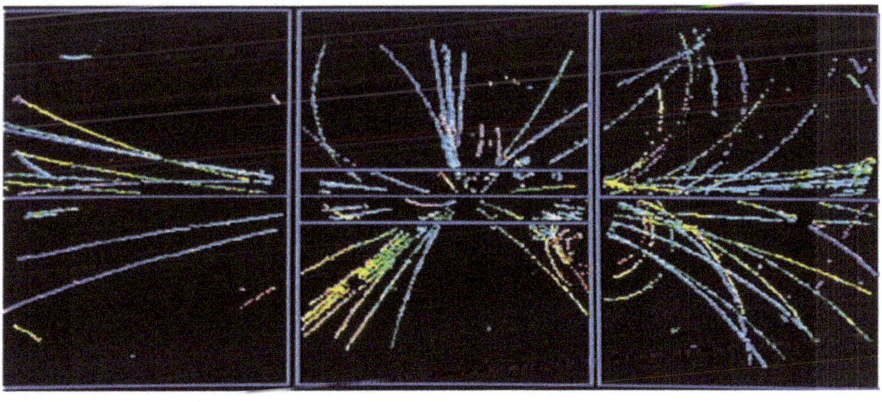

Figure 9.6. Computer display of the outcomes of a proton-antiproton collision in the UA1 drift chambers.

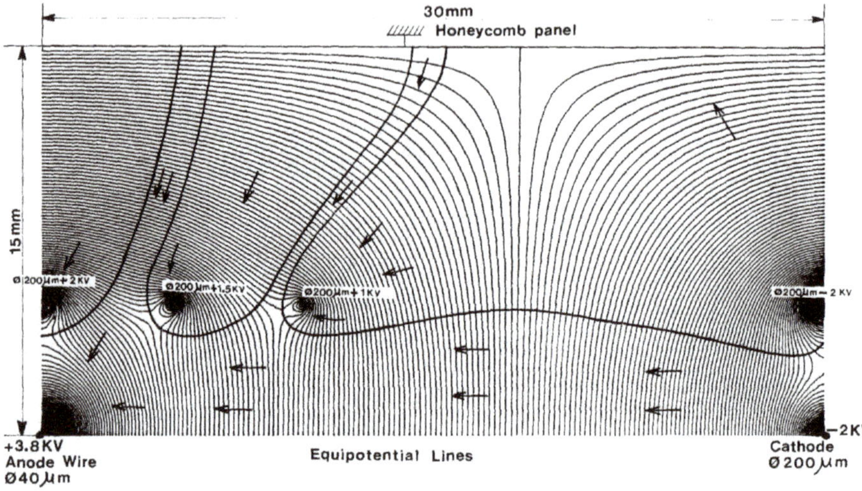

Figure 9.7. Field structure in a cell of the large drift chambers for the WA1 experiment at CERN (Marel *et al.*, 1977).

Figure 9.8. The WA1 neutrino detector at CERN (Picture CERN).

iron, scintillation counters and drift chambers of hexagonal shape, four metres in diameter, for position measurements. Two-dimensional localization is achieved mounting triplets of chambers with wires at 60 degrees. Charged particles created by neutrino interactions in the iron plates were localized with about one millimetre accuracy and identified by magnetic analysis. The results of the experiment permitted significant advances in the study of neutral currents, neutrino oscillations and quantum chromodynamics.

9.2 Cylindrical Wire Drift Chambers

In the seventies, drift chambers with cylindrical geometries of various designs have been built, better suited for installation in the solenoids used for magnetic tracks analysis; Figure 9.9 is a close-up view of the wire structure in the Time Expansion Chamber (TEC), operated in gas and field conditions optimized to achieve best radial position accuracy (Walenta, 1979); Figure 9.10 displays an event recorded with the TEC of experiment L3, one

Figure 9.9. Close-up of the wire structure in the Time Expansion Chamber (Picture CERN).

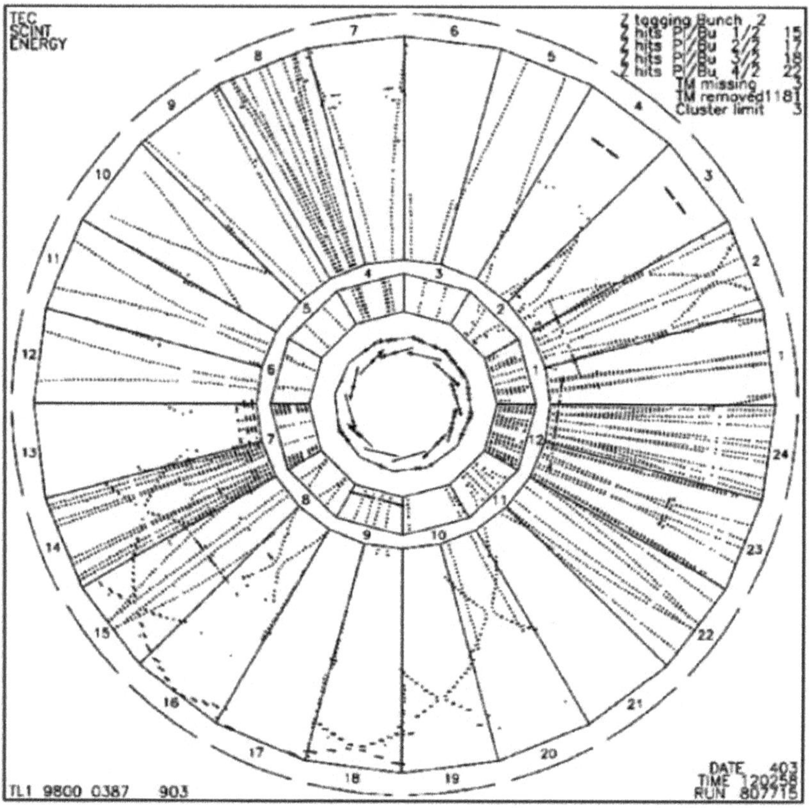

Figure 9.10. An event recorded with the TEC of the L3 experiment at CERN (Anderhub *et al.*, 2003).

of the detectors at CERN's Large Electron Positron Collider (LEP) operated between the years 1989 and 2000 (Anderhub *et al.*, 2003).

Optimized for the detection of high local track density (jets), the structure of central JET chamber of the JADE detector atthe e^+-e^- storage ring PETRA is shownin Figure 9.11. Figure 9.12 is an example of recorded event with two jets of charged particles (Drumm *et al.*, 1980). The sensitive volume is subdivided in 24 segments; in each sectors, a central array of anode wires alternates with field wires; the right-left ambiguity is resolved by staggering the anodes alternatively to the left and to the right of the median plane. The longitudinal coordinate is deduced

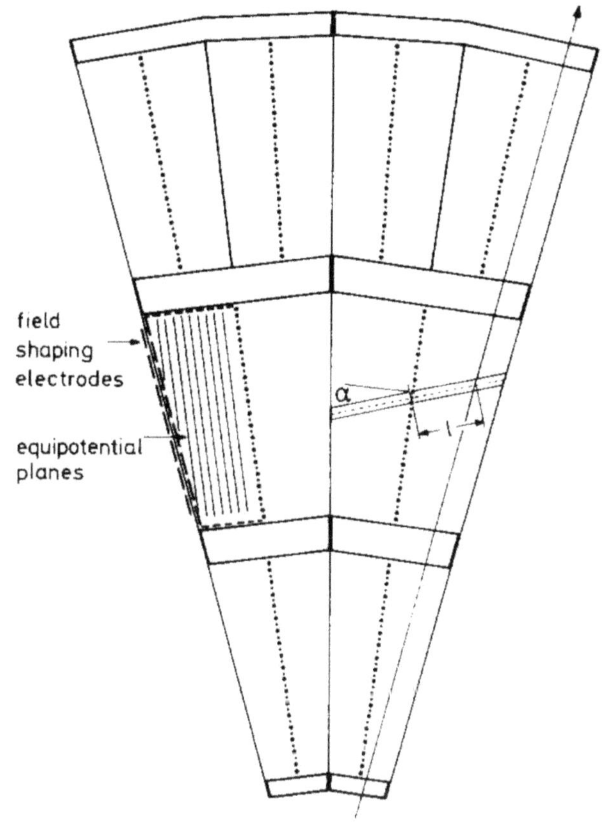

Figure 9.11. Design of the JADE JET Chamber (Drumm *et al.*, 1980).

from the ratio of charges recorded at the two ends of each wire.A statistical analysis of the differential energy loss, sampled on the wires, and a measurement of the track bending in magnetic field help identifying the charged tracks.

Similar in design, the JET drift chamber of the OPAL experiment at LEP, operated from 1989 until decommissioning of the collider, was designed by two concentric cylindrical modules 4 m long and with an outer diameter of 3.7 m (OPAL_Collaboration, 1991).

Analogous in design, the MARK II Cylindrical Wire Drift Chamber operated from 1997 at the Stanford Linear Accelerator, mainly to study

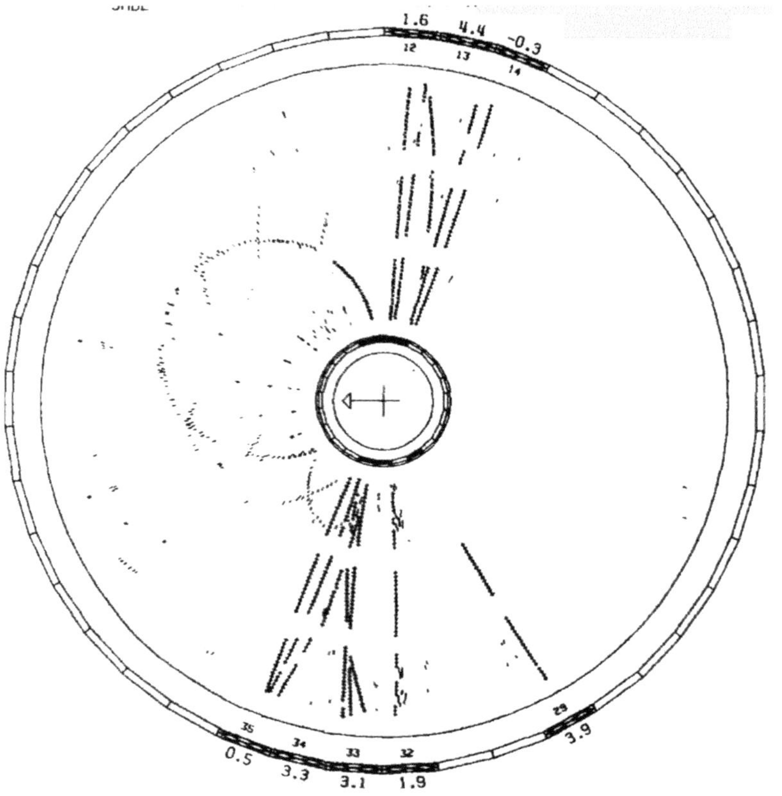

Figure 9.12. An event recorded with the JADE JET chamber (Drumm *et al.*, 1980).

the decay modes of the Z_0 boson (Abrams *et al.*, 1989); Figure 9.13 is a view of the detectors during the stringing of the sense and field wires. The cells were arranged in 12 concentric cylindrical layers, alternating between wires parallel to the cylinder axis or inclined at approximately ± 4° to the axis to provide stereo information.

Cylindrical drift chamber has been operated in other laboratories from the early nineteen ninety, as for example TOPAZ and VENUS at the storage ring TRISTAN of the National Laboratory for High Energy Physics (KEK, Japan). With their time resolution and memory, typically between tens and a few hundred nanoseconds, the detectors are suited for use at rather high collision rates.

Figure 9.13. The MARK II cylindrical drift chamber during construction (Picture SLAC).

9.3 Large Volume and Time Projection Chambers

The tracking capabilities of wire drift chambers are blighted by the complex fabrication and relative fragility. Aiming specifically at charged particle identification up to ~100 GeV, large volume drift chambers named ISIS (Identification of Secondary Particles by Ionisation Sampling) were built at the Oxford Nuclear Physics Laboratory (Allison *et al.*, 1974). As seen in the picture of Figure 9.14, the detectors consisted in a large gas vessel with field-shaping electrodes on the side walls, two cathode planes at each end and a MWPC plane in the middle of a 2 m thick drift volume,

Figure 9.14. The ISIS2 large volume drift chamber (Allison *et al.*, 1984).

and operated with an argon-carbon dioxide mixture at one bar of pressure. Figure 9.15 is an example of multi-prong event recorded with ISIS2 in the European Hybrid Spectrometer at CERN's Super Proton Synchrotron (SPS) (Allison *et al.*, 1984); the brightness of tracks is representative of the local ionization density. A measurement of the differential energy loss, followed by a statistical analysis of the data provides identification of electrons, pions and protons in the quoted momentum range.

The Time Projection Chamber (TPC) designed by David Nygren in 1978 provides exceptional performances for tracking and particle identification (Nygren and Marx, 1978); for a review of the TPC history and development see (Nygren, 2018). The device consists in a large gas vessel, where the ionization trails released by charged particles are drifted by the electric field to an end-cap array of MPPC-based detectors recording,

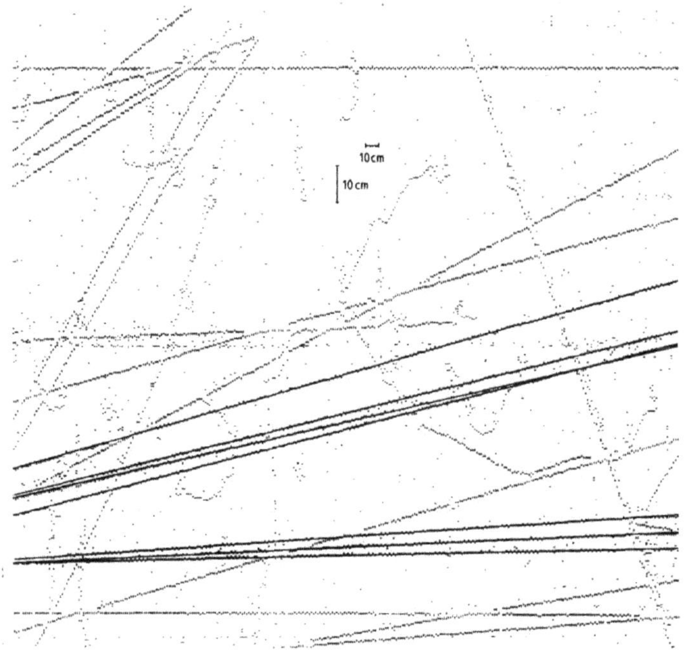

Figure 9.15. A multi-track event recorded with the ISIS detector (Allison *et al.*, 1984).

within short time slices, the succeeding segments of ionization charges, hence the name. Sets of wires or strips at suitable potentials, capped by a cathode plane, surround the vessel providing a uniform field in the sensitive volume. Within each time slice, the anode wire signals and the profile of charge induced on rows of facing cathode pads are recorded, permitting to measure the projected position of the track, its drift time and differential energy loss, Figure 9.16.

Figure 9.17 shows the original design of the PEP-4 TPC, built at BNL and operated at SLAC to detect the outcomes of electron-positron collisions (Fancher *et al.*, 1979). Cylindrical in shape, the detector was installed inside a 4 kG solenoid for momentum analysis, and consisted in two back-to back drift volumes, 1 m long each, sharing a cathode plane in the centre, and two MPWC-equipped end caps for tracks recording. To optimize the particle identification properties, exploiting the strong reduction of transverse electron diffusion in magnetic field, the TPC was

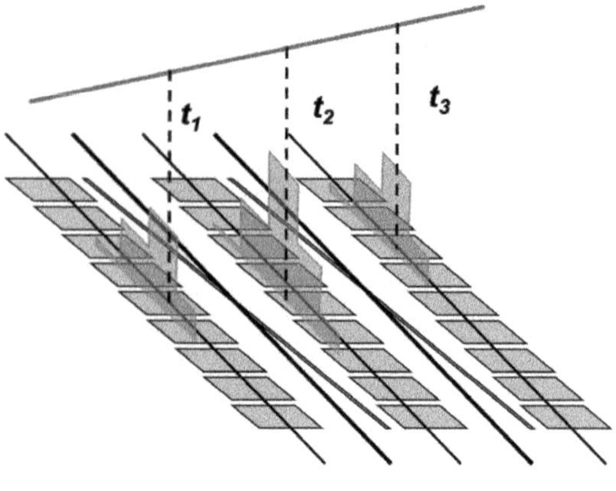

Figure 9.16. Signals readout on the TPC end-cap MWPC.

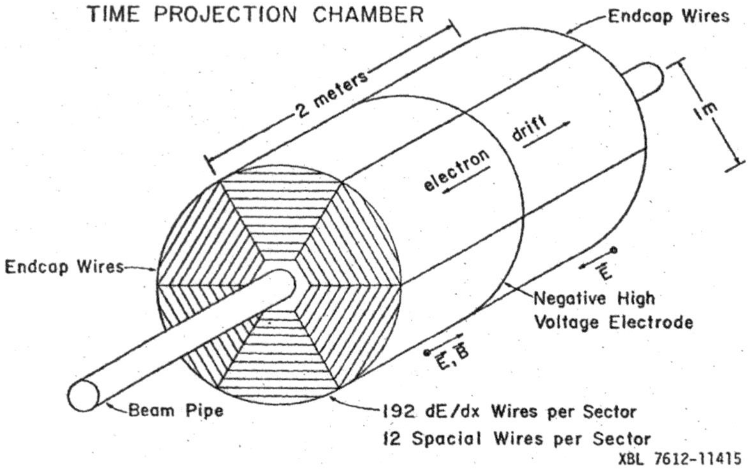

Figure 9.17. Design of the PEP-4 TPC (D. Nygren and J. Marx, 1978).

operated in an argon-methane mixture at a pressure of 8.5 atmospheres.[2] The detector had six independent MWPC sectors each side, with 183 sense wires at 4 mm pitch; fifteen of the sense wires in each sector had the cathode plane locally segmented into strips of square 7.0 × 7.5 mm^2 pads, (Aihara *et al.*, 1983). The picture in Figure 9.18 shows one of the MWPC modules, before assembly on the TPC end cap.

[2]In David Nigren's own words, simulations indicated the optimum pressure to be between 8 and 9 atmospheres; the choice was suggested by Fellini's movie "8 ½".

Figure 9.18. A MWPC module before assembly (courtesy D. Nygren).

The simultaneous measurement of the projected coordinates and the drift time provides a complete three-dimensional image of events, as shown in Figure 9.19. The ionization charge recorded on each segment of tracks, followed by a statistical analysis, provides the most probable values of energy loss; plotted as a function of the particle's momentum, deduced from the track's bending in magnetic field, the average differential energy loss permits identification of the particle masses in a wide range of momenta, Figure 9.20.

Cylindrical time projection chambers of similar design have been built to instrument many experiments worldwide. The ALEPH and DELPHI TPCs operated at CERN's large electron-positron storage rings (LEP) between 1988 and 2000; both detectors were made of two back-to-back cylindrical drift volumes fitting inside a superconducting solenoidal magnet and separated by a common high-voltage electrode; arrays of trapezoidal MWPCs at each end served as amplifying and charge recording detectors. Figure 9.21 shows the structure of the ALEPH TPC, 360 cm in diameter and with 470 cm active length (Decamp *et al.*, 1990); Figure 9.22 is a front view of the central tracking complex before installation of the forward endcap calorimeters and muon detectors. Nobel laureate Jack Steinberger, leader of the collaboration, stands third from the left.

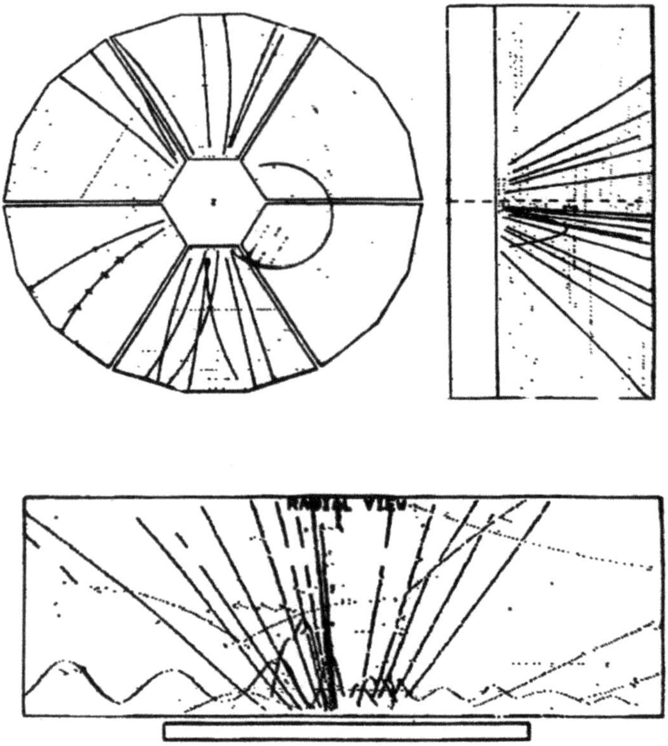

Figure 9.19. Projected three-dimensional reconstruction of an event recorded with the LBL TPC (courtesy D. Nygren).

The picture in Figure 9.23 displays an event reconstructed in the relatively smaller DELPHI TPC, surrounding the silicon strip vertex detectors described in Section 6.1.

Dedicated to the study of ion collisions at Brookhaven's Relativistic Heavy Ions Collider, the Solenoidal Tracker at RHIC (STAR) includes a Time Projection Chamber optimized for the recording of high multiplicity events; with its 400 cm diameter,when it was commissioned in 1999 it was the largest of its kind. Figure 9.24 is the impressive display of the outcomes of a collision between two Au ions at 30 GeV per beam, resulting in thousands of charged particles prongs (Anderson *et al.*, 2003). The experiment contributed to the study the quark-gluon plasma properties.

Approximately 5 m long and having 90 m^3 of active volume, the ALICE time projection chamber, Figures 9.25 and 9.26, has been

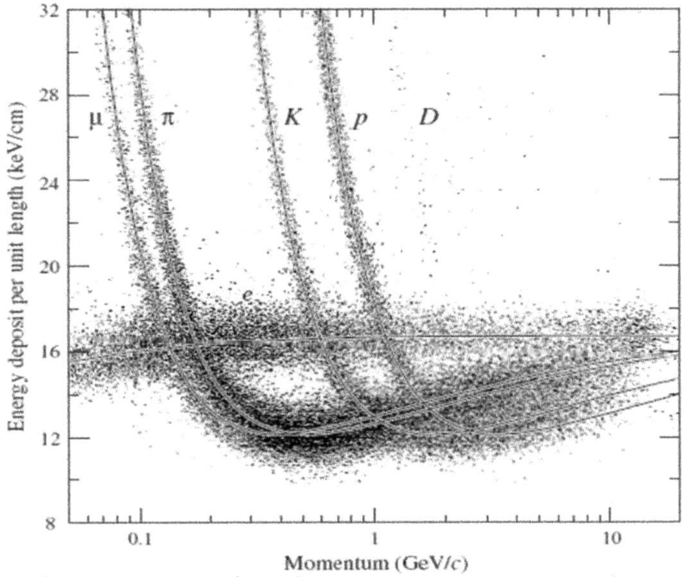

Figure 9.20. Particle identification obtained from the measured differential energy loss as a function of momentum (courtesy D. Nygren).

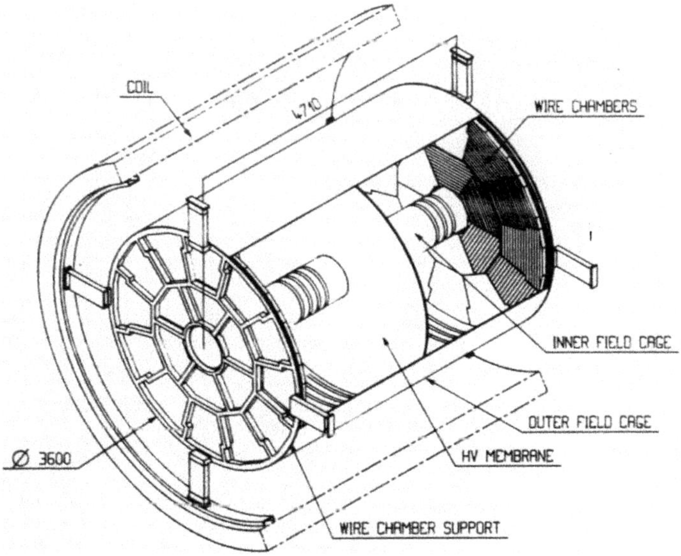

Figure 9.21. Schematics of the ALEPH time projection chamber (Decamp *et al.*, 1990).

Figure 9.22. Front view of the ALEPH central detector complex. The TPC MWPC array is surrounded by the solenoidal magnet coil and the barrel hadronic calorimeters (Picture CERN).

operational from 2008 at CERN's LHC (Alme *et al.*, 2010); in 2020, in order to cope with the planned increase of luminosity of the collider, the TPC was upgraded replacing the MWPCs with a detection and readout system based on the Gas Electron Multiplier (GEM), see Section 11.4 (Hauer, 2022).

Rectangular TPCs, better suited to fixed target experiments, have been operated in various laboratories worldwide; an example is the NA49 complex at CERN, continued as NA61-SHINE (Abgrall *et al.*, 2014). The

Figure 9.23. A two-jets event recorded in the DELPHI TPC, surrounding the silicon microstrip vertex detector (Picture CERN).

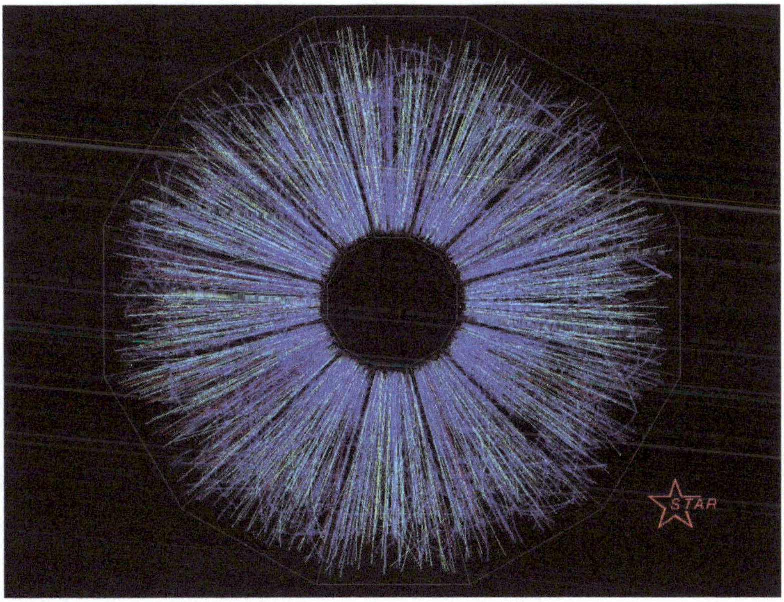

Figure 9.24. An Au-Au collision in the STAR TPC at RHIC (Anderson *et al.*, 2003).

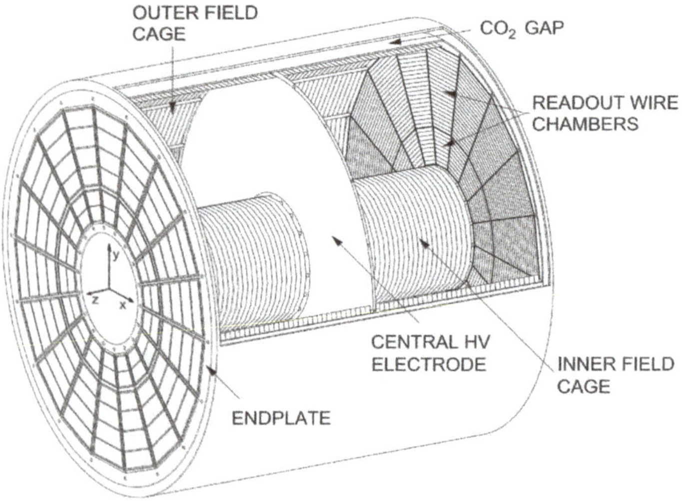

Figure 9.25. Schematics of the ALICE TPC (Alme *et al.*, 2010).

Figure 9.26. Front view of the ALICE TPC; sitting in the centre Peter Glässel, TPC technical coordinator (Picture CERN).

experiments are designed to study hadron production in proton-proton, proton-ion and ion-ion interactions at very high energies. Figure 9.27 is the impressive display of a Pb-Pb interaction at 158 GeV/nucleon, yielding thousands of charged particles. The experiment assembles within a pair of large dipole magnets sets of vertex and main TPC modules, varying in size from 200 to 390 cm drift length; the blue outlines in the display mark the position of the TPC modules.

Another example is the near neutrino detector at the Tokay-to-Kamioka (T2K) experiment in Japan (Abgrall *et al.*, 2011), aiming at the study of neutrino oscillations. A set of off-axis TPCs was used to assess the quality of the neutrino beam; MICROMEGAS micro-pattern detectors record the charged prongs of interactions Figure 9.28 The experiment has been upgraded in 2020 with resistive MICROMEGAS devices, capable to withstand higher rates and providing better position

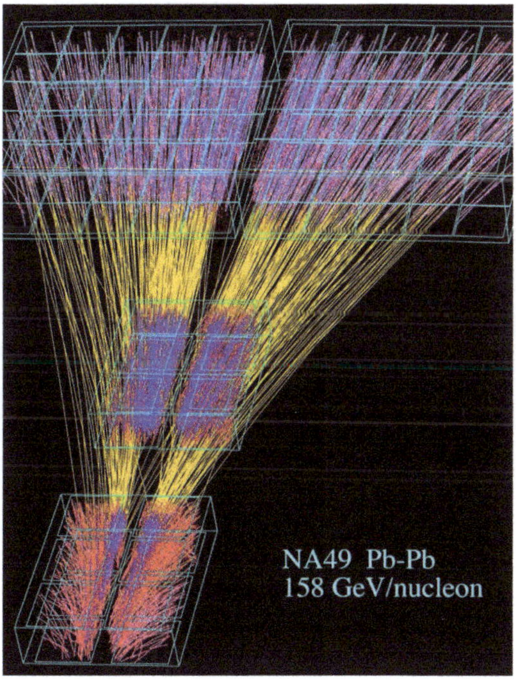

Figure 9.27. A Pb-Pb collision recorded with the TPC complex of experiment NA39 (Picture CERN).

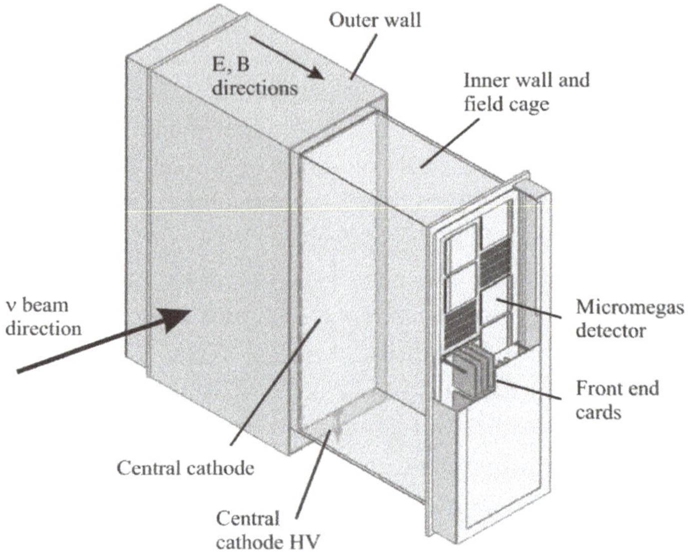

Figure 9.28. A module of the T2K TPC detector (Abgrall *et al.*, 2011).

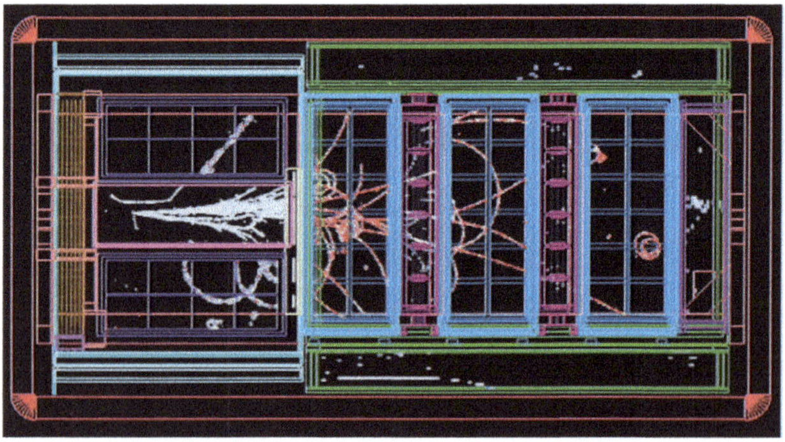

Figure 9.29. A neutrino interaction recorded with the upgraded T2K experiment (Kutter, 2023).

accuracy (see Chapter 11); neutrinos interact in a 3D scintillating fibre target, followed by three forward MICROMEGAS TPCs. Figure 9.29 shows one of the first neutrino interactions recorded with the upgraded detector (Kutter, 2023).

9.4 Limited Streamer Tubes

While providing exceptional tracking performances, the devices described in the previous sections are rather complex and unsuitable for very large volume coverages, as those required in the detection of muons, neutrinos, cosmic rays or rare events. Introduced by Enzo Iarocci in 1978, limited streamer tubes allow to build very large arrays of position-sensitive detectors using simple and cheap manufacturing technologies (Battistoni *et al.*, 1979). As observed by the authors, single- or multi-wire counter arrays with thick (50–200 μm) anode wires operated in heavily quenched gas mixtures and with high-resistivity cathodes permit to reach a regime of streamer propagation, limited in length to a fraction of the anode-cathode distance.

Figure 9.30. Plastic limited streamer tubes during manufacturing (Picture CERN).

While having moderate position accuracy and low-rate capability, the limited streamer operation provides large and saturated pulses particularly convenient to instrument huge numbers of channels.

One of the first large scale implementation of limited streamer tubes was for the Nuclear Stability Experiment (NUSEX) detector, installed in the tunnel under the Mont Blanc and aiming at the study of the nucleon stability (Battistoni *et al.*, 1986). The detector was a digital tracking calorimeter, consisting of a sandwich of horizontal iron plates interleaved with arrays of plastic streamer tubes 3.5 m long and 9×9 mm^2 in cross section. The tubes are assembled in groups in an extruded PVC plastic box, with the internal walls coated with graphite; plastic holders help keeping the wires in place. Figure 9.30 shows several modules during construction.

With a total number of tubes close to 50,000, the experiment operated between 1982 and 1988 setting a limit to the proton decay time of about 10^{30} years; it was also utilised to continuously monitor high-energy muons and neutrinos from stellar sources.

Large arrays of plastic streamer tubes, interleaved with heavy metal converters, are universally employed in particle physics experimental set-ups to for hadron calorimetry (Iarocci, 1983). In astrophysics, they permit to cover the large detection areas required for the study of extended cosmic ray showers, see for example reference (Antoni *et al.*, 2004).

Chapter 10

Resistive Plate Chambers

Parallel plate counters, in which charge multiplication occurs in uniform electric fields, have been used as integral flux detectors, but only sporadically for single event detection due to their tendency to discharge at high gains. A substantial improvement in operating stability was achieved by Rinaldo Santonico and Roberto Cardarelli, from Rome University, with the development in the early 1980 the Resistive Plate Chamber (RPC) using as electrodes high-resistivity plates and heavily quenched filling gases; this permitted to obtain full efficiency and very good time resolution in the detection of minimum ionizing particles, albeit at the cost of a restrained rate capability (Santonico and Cardarelli, 1981). The RPC structure is shown in Figure 10.1; several variants of this basic design have been developed. Two high resistivity electrodes, coated on the outer side with a moderate resistivity layer to apply the operating voltage and transparent to fast pulses, encompass a thin gas gap; padded external electrodes sense the signals providing time and position information. With a suitable choice of parameters, the detectors can be operated either in an avalanche or in a limited streamer mode: on the occurrence of a large local surge, the current in the high resistivity electrode results in a drop of electric field, preventing the transition to a discharge.

The detector performances, namely efficiency and time resolution, can be improved assembling two single-gap RPCs back-to-back with a common signal pickup electrode in the centre, Figure 10.2.

Aiming at the realization of very large detection areas, RPCs have been realized with a wide range of materials having resistivities in the

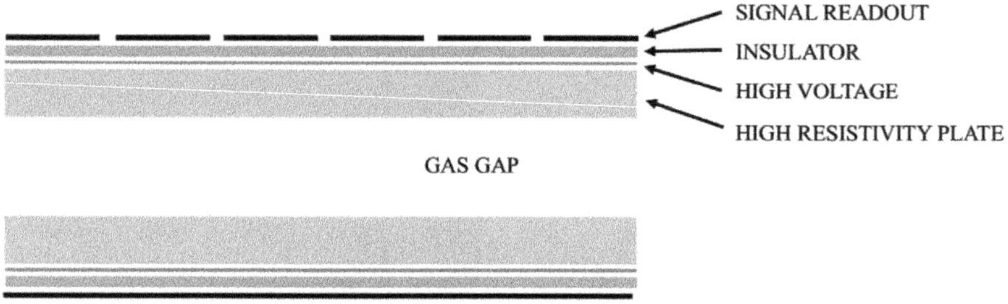

Figure 10.1. The Resistive Plate Chamber.

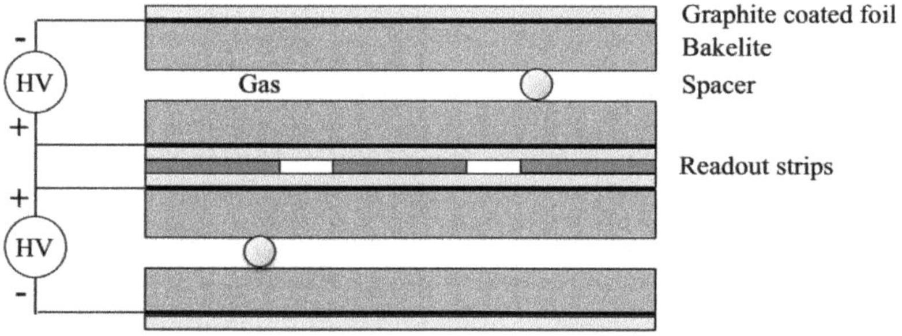

Figure 10.2. Double-gap RPC with a common readout electrode.

range 10^{10}–10^{12} Ωcm such as graphite-coated insulators, bakelite, melamine, resistive glass. The required surface uniformity is ensured by thorough polishing, often limited in quality however by the roughness of the material; for bakelite RPC, a widespread solution to smooth the electrodes was to coat them with linseed oil. While ensuring stable long-term operation, local imperfections and chemical reactions with moisture induce a process of performance deterioration referred to as ageing, subject of extensive studies (Abbrescia *et al.*, 1997).

The RPCs evolution, manufacturing methods and operating characteristics, as well as their use in experiments are described in the comprehensive book by Marcello Abbrescia, Vladimir Peskov and Paulo Fonte (2018) and in a wide number of publications. Model calculations of the signal formation in resistive detectors are discussed, for example in Ref (Fonte, 2013).

Large arrays of RPC plates have and still are used in particle physics experiments: BaBar at SLAC, HADES at GSI Darmstadt; L3, ALICE,

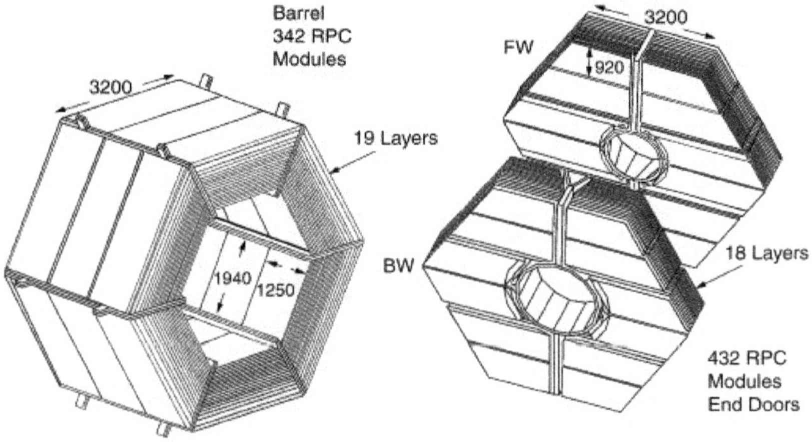

Figure 10.3. Barrel, forward and backward RPC modules of the BaBar experiment at SLAC (Anulli *et al.*, 2002). Dimensions in millimetres.

ATLAS, CMS at CERN and many others; they generally instrument the outer shells of the experiments to detect muons. As an example, Figure 10.3 displays the barrel, forward and backward systems of the BaBar experiment, including close to 800 RPC modules (Anulli *et al.*, 2002). The picture in Figure 10.4 shows the CMS forward RPC muon detector during assembly; similar systems are deployed in other LHC experiments.

The ARGO–YBJ experiment (Astrophysical Radiation with Ground-based Observatory at YangBaJing) at an altitude of 4300 m above the sea level includes approximately 2000 large size RPCs covering a surface close to 10,000 m^2. Its purpose is the detection of large atmospheric showers generated in the atmosphere by energetic cosmic rays. The plates were produced in Italy using the bakelite-linseed oil manufacturing technique developed for BaBar; the initial performances of the system are described in Ref. (Camarri, 2007). Figure 10.5 is an inside view of the ARGO–YBJ hall with RPCs deployed on the floor (Cao, 2021).

The experiment commenced in 2007 and continued for many years it provided vast amount of results relevant for γ-rays astrophysics, for the search of large-scale anisotropy at TeV energies and proton–air interaction cross-sections; for a review see (Aielli *et al.*, 2012; Di Sciascio, 2014).

Figure 10.4. CMS forward RPC muon detector (Picture CERN).

Figure 10.5. A view of the ARGO–YBJ hall (Cao, 2021).

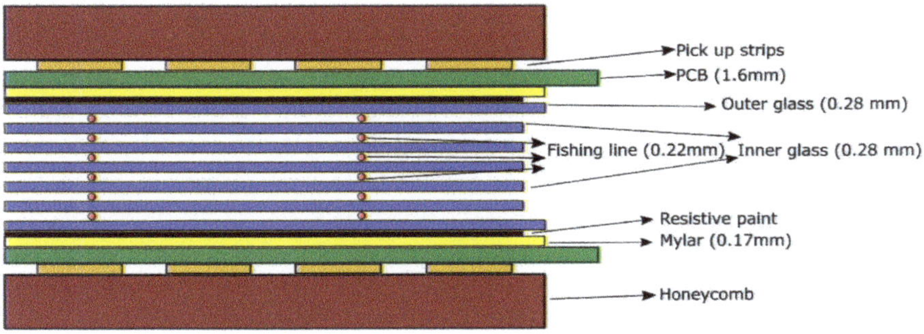

Figure 10.6. A six-gap MRPC (Liu *et al.*, 2018).

Developed by Crispin Williams and his group in 1995, the multi-gap resistive plate chamber (MRPC) achieves very good time resolutions and high rate capability (Cerron Zeballos *et al.*, 1996). Figure 10.6 shows schematically a MRPC stack with six 0.16 mm gas gaps between thin glass plates; the uniformity of the response is ensured by insulating fishing lines spacers. Light honeycomb plates on each side provide mechanical rigidity (Liu *et al.*, 2018). With a six-gap MRPC a time resolution around 70 ps has been achieved for minimum ionizing particles, improving to ~30 ps for a 20-gap multiple stack. A large MRPC-based system has been operated in the ALICE experiment at CERN from 2009 providing particle identification through the measurement of Time-of-Flight with around 160 ps resolution (Akindinov *et al.*, 2012). Similar systems equip other setups: STAR at RHIC, HADES and the Compressed Baryonic Matter (CBM) experiments at Darmstadt. For a recent review of MRPC-based Time-of-Flight systems see Ref. (Wang and Yu, 2021).

Chapter 11

Micro-Pattern Gaseous Detectors

11.1 Microstrip Gas Counter (MSGC)

Despite their remarkable performances and the impact on particle physics experimentation, wire-based gaseous detectors suffer from several shortcomings: they are rather fragile, and instabilities due to electrostatic forces limit the distance and length of unsupported wires. A more fundamental limitation results from the slow motion of positive ions generated in the avalanche process, due to the time required to neutralize at the cathodes: at high radiation fluxes, the accumulation of positive space charge causes a decrease of the electric field around the wires, and a reduction of gain at particle fluxes exceeding ~10^4 $mm^{-2}s^{-1}$.

An innovative detector, developed in 1988 by Anton Oed at the Institute Laue Langevin (ILL) in Grenoble, permits to overcome some of these limitations. Named Microstrip Gas Counter (MSGC), the device consists in a set of thin parallel metallic strips engraved on an insulating substrate, alternatively connected as thin anodes and wider cathodes, Figure 11.1 (Oed, 1988). On application of a difference of potentials between adjacent strips, and with the overlaying gas volume delimited by a drift electrode, the electric field lines converge to the anodes creating a field structure evocative of the one of wire chambers: electrons released by ionization in the gas volume drift towards the anode, multiply in avalanche and are recorded. The majority of the positive ions created in the process are swiftly collected and neutralized on the neighbouring cathode strips, minimizing the space-charge field distortions; added to the narrow

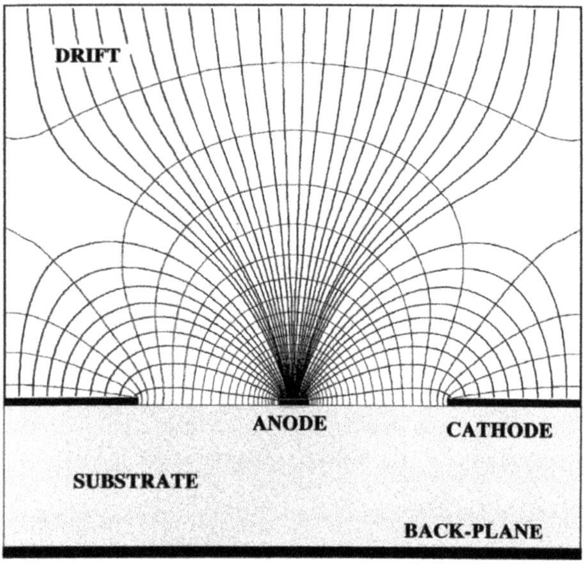

Figure 11.1. Structure and electric field lines in the MSGC.

pitch, this results in a two orders of magnitude increase of the rate capability of the detector, that can reach $10^6\,mm^{-2}s^{-1}$.

Using photolithographic technologies, the distance between strips can be a fraction of mm; a few hundred microns pitch chromium strips engraved on glass are a common choice. Figure 11.2 is a close-up of an MSGC plate with 100 μm wide anode strips at 200 μm pitch.

Dedicated to the detection of crystal neutrons diffraction, the D20 diffractometer at ILL, shown in Figure 11.3, was an assembly of 48 MSGC modules mounted with semi-circular shape. Using an ^3He–CF_4 gas mixture, the spectrometer started operating in 1997 but after about two years of use suffered some performance deteriorations and was replaced by an improved system named D4C (Clergeau *et al.*, 2001).

The ILL experience with MSGCs revealed their major weakness: the long-term operation is altered by the presence of even very small amounts of pollutants, creating deposits on the thin anodes and affecting the gain. The structures are also easily damaged by discharges, particularly if operated at the gains required to detect small ionization yields. Plans to use the devices to equip the large detectors used in particle physics were therefore discontinued.

Figure 11.2. Microscope view of an MSGC plate (Picture CERN).

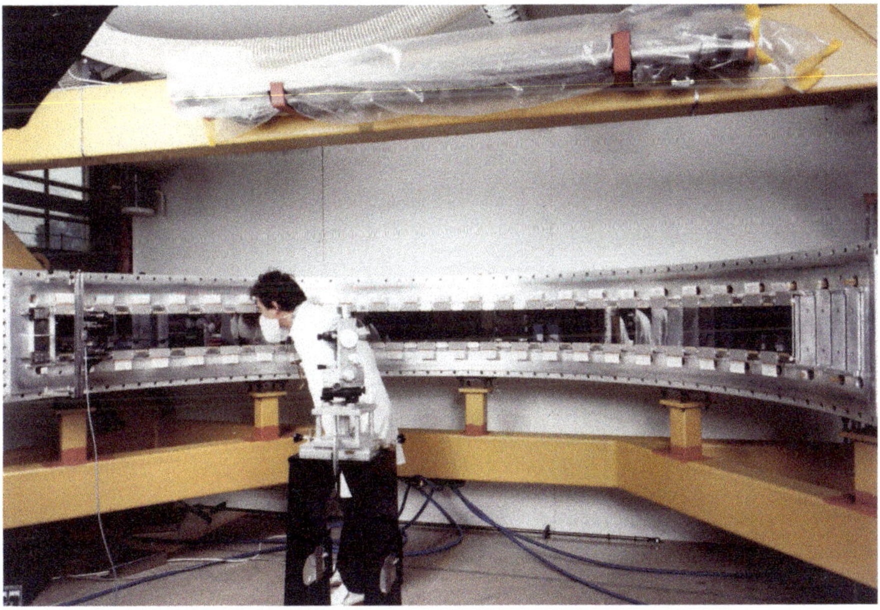

Figure 11.3. The D20 neutron diffraction spectrometer at ILL Grenoble.

11.2 Early Micro-Pattern Detectors Design

The large collaborative effort aiming at finding a solution of the problems encountered with MSGCs resulted in the elaboration of new technologies to improve the photolithographic processing and control the surface conductivity of the insulating supports; the know-how was exploited to develop a new generations of micro-pattern devices of various design, see for example Chapter 3 in the author's book on Micro-Pattern Gaseous detectors (Basic Bibliography). Often too difficult to manufacture or having other limitations, most of these designs remained at the prototype stage. An exception is the micro-groove detector (MGD) shown in Figure 11.4, with anode strips buried in a narrow channel etched on an insulator, with cathode strips on the top; the increased distance between electrodes acts as effective protection against discharges (Bellazzini *et al.*, 1999).

An improved design of the structure, realized on a polymer coated with high-resistivity diamond-like carbon (DLC) layers, named micro-resistive WELL (μ-RWELL) permits to reach high stable gains while providing immunity against discharges, Figure 11.5 (Bencivenni *et al.*, 2017). Addition of grounded current return strips at intervals permit to preserve efficiency at radiation fluxes well above 1 MHz cm^{-2}; Figure 11.6

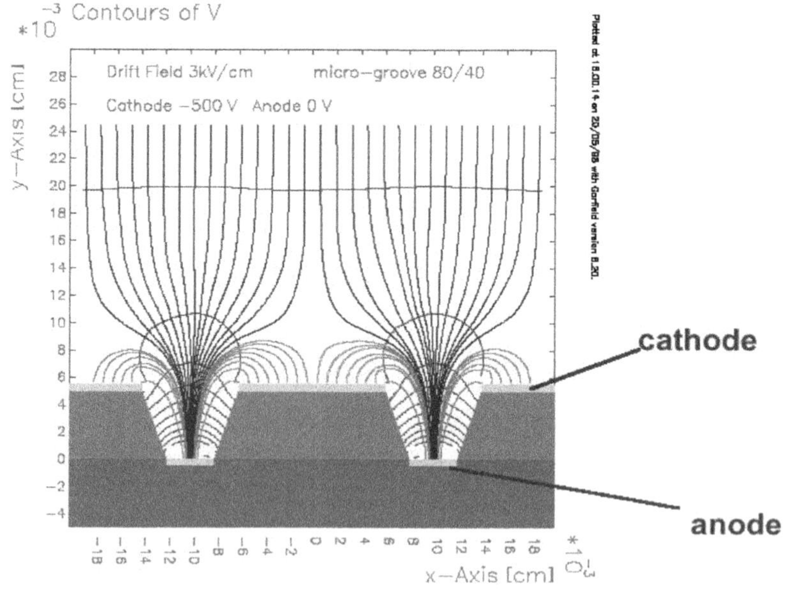

Figure 11.4. The micro-groove detector (Bellazzini *et al.*, 1999).

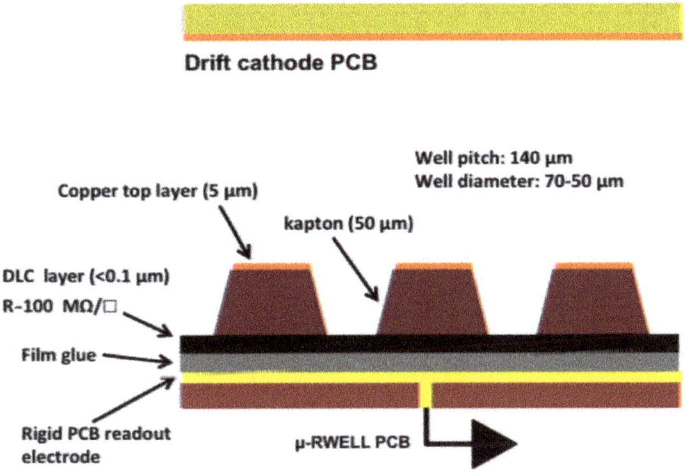

Figure 11.5. The resistive micro-WELL detector (Bencivenni *et al.*, 2017).

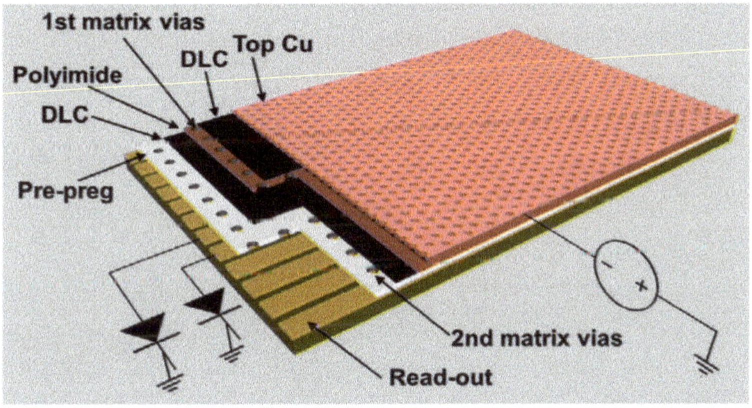

Figure 11.6. The double-resistive micro-WELL (Bencivenni *et al.*, 2020).

shows an advanced design combining in a single artwork two resistive layers, through-connections (vias) and readout strips (Bencivenni *et al.*, 2020). While rather complex to manufacture, the device has the advantage to provide the high resolution, high-rate capability of other MPGDs with a single amplification stage, and has been proposed for CERN's LHCb muon detector upgrade (Paolucci, 2022).

11.3 Micro-Mesh Gaseous Counter (MICROMEGAS)

Introduced in 1996 by Ioanis Giomataris, the micro-mesh gaseous counter (MICROMEGAS) exploits a property of the multiplication coefficient in narrow gas gaps to saturate at high electric fields, thus permitting more stable operation at large gains. As shown in Figure 11.7, the detector has a region of high field separated by a metallic mesh from a low field gap; electrons released by ionization in the upper volume drift into the high field region where charge multiplication occurs. With proper choice of conditions and multiplication gaps around 100 μm, proportional gains above 10^4 can be reached safely permitting to detect and localize minimum ionizing particles (Giomataris *et al.*, 1996).

Various techniques have been developed to ensure the multiplying gap thickness over large areas; the most current, the so-called bulk MICROMEGAS, consists in the lamination over an anode plane of a thin photoresistive film and of a metallic micromesh. After curing, the

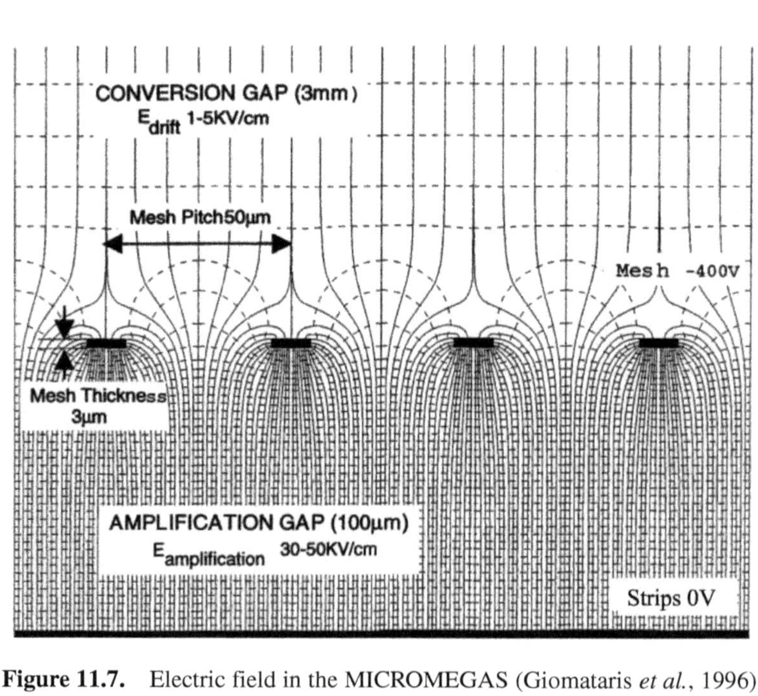

Figure 11.7. Electric field in the MICROMEGAS (Giomataris *et al.*, 1996).

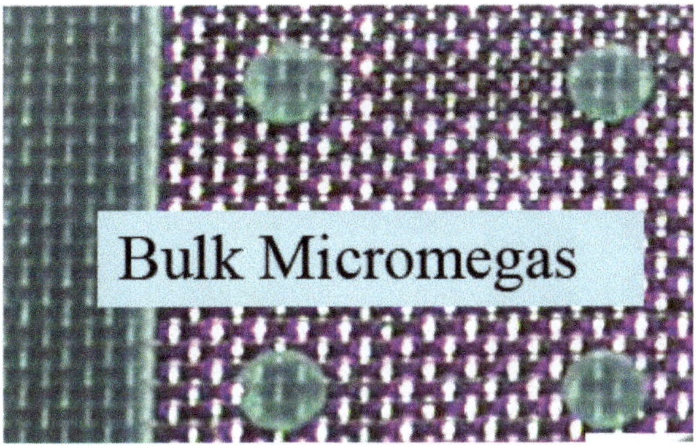

Figure 11.8. Close-up of the gap-restoring insulating pillars in the bulk MICROMEGAS (Giomataris *et al.*, 2006).

insulator is removed by a photolithographic process through a mask, leaving narrow pillars connecting anode and mesh, Figure 11.8 (Giomataris *et al.*, 2006).

Thanks to the favorable statistics of charge multiplication in uniform fields, the MICROMEGAS has excellent energy resolution for soft X-rays, 11% at 8 keV.

A medium-scale system of MICROMEGAS detectors has been built and operated for many years in the COMPASS spectrometer at CERN. The chambers, with 40×40 cm^2 active area, have projective strips readout and are mounted in doublets rotated by 45° thus providing 2D track directions, Figure 11.9 (Bernet *et al.*, 2005). For perpendicular particle tracks, recording of the charge sharing between adjacent signal strips provides localization accuracies around 50 μm.

Several types of resistive MICROMEGAS have been developed to eliminate the discharge problems observed when operating the detector in harsh radiation environments. Figure 11.10 shows one of the possible schemes, with the anode implemented with high-resistivity strips; the signals are collected by capacitive coupling on the readout strips (Alexopoulos *et al.*, 2019). This solution has been adopted for the ATLAS muon spectrometer upgrade; Figure 11.11 shows one of the modules in construction and Figure 11.12 the new muon small wheel during assembly.

Figure 11.9. A fully equipped MICROMEGAS detector in CERN's COMPASS spectrometer (Bernet *et al.*, 2005).

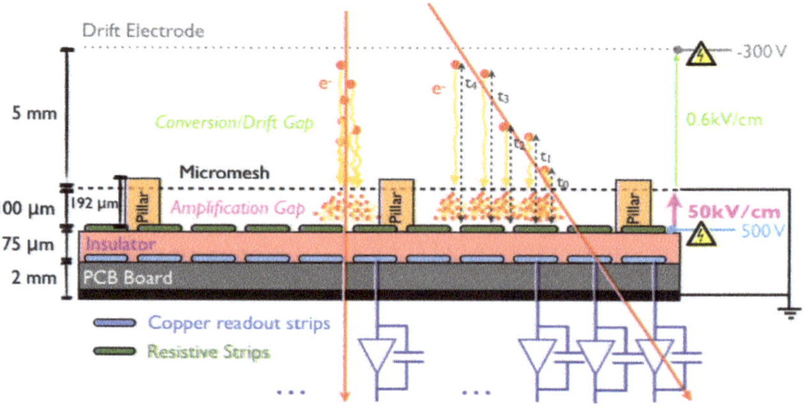

Figure 11.10. The resistive MICROMEGAS; the anode strips, with typically 10 MΩ cm^{-1}, prevent the onset of discharges (Alexopoulos *et al.*, 2019).

Figure 11.11. A resistive MICROMEGAS module for the ATLAS muon spectrometer upgrade (Alexopoulos *et al.*, 2020).

Figure 11.12. The ATLAS small wheel muon detector upgrade during assembly (Picture CERN).

The MICROMEGAS detector technology has found numerous applications in particle and nuclear physics, astrophysics and other fields; for a comprehensive review see (Attié *et al.*, 2021).

11.4 Gas Electron Multiplier (GEM)

Introduced by Fabio Sauli in 1997, the Gas Electron Multiplier (GEM) performs amplification of ionization with the transfer of a sizeable fraction of the avalanche electrons to cascaded amplifying or collecting electrodes, thus permitting to safely reach large gains (Sauli, 1997).

A GEM electrode consists in a thin polymer foil, metal-clad on both sides and pierced by a high density of holes, typically 50–100 mm^{-2}, using high-resolution photolithography (Figure 11.13). On application of a difference of potential between the top and bottom electrodes, the electric field in the region of the holes can be increased to the value required for charge multiplication (Figure 11.14). With a proper choice of external fields, ionization electrons released in the upper volume drift into the high-field region within the holes and multiply in avalanche; a field-dependent fraction of the electrons escapes the holes into the lower

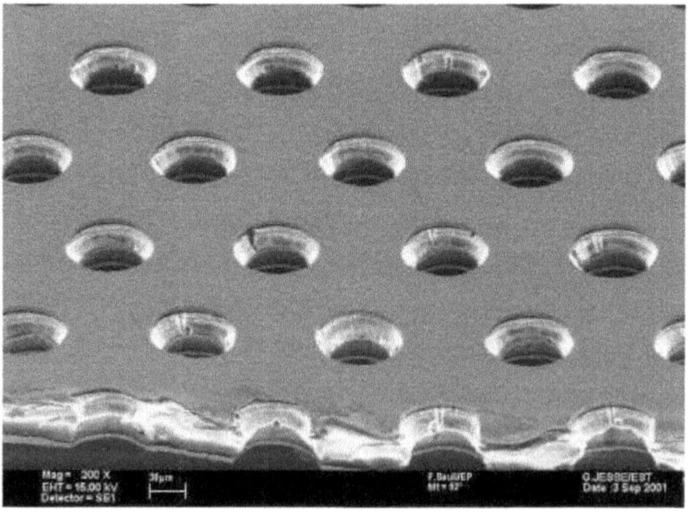

Figure 11.13. Close-up of a GEM electrode (Picture CERN).

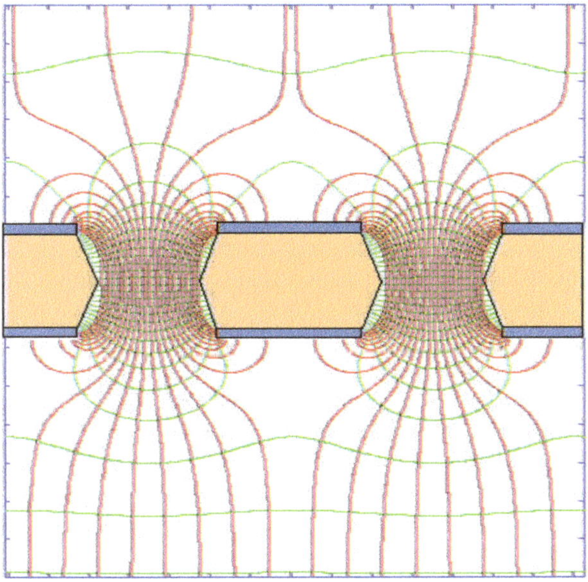

Figure 11.14. Electric field lines and equipotential in the GEM holes.

transfer gap, where it can be collected or further amplified. Localization is then performed on a collecting anode, recording the charge on finely pitched strips or pads. Figure 11.15 shows schematically a single GEM detector assembly, with a charge multiplying electrode and perpendicular strips on the anode providing bi-dimensional coordinate information. The pulse detected on the GEM bottom electrode can be used to provide a trigger and energy discrimination. Several GEM electrodes can be mounted in cascade, providing large amplification factors substantially reducing the probability of discharge, a common choice being the Triple-GEM (Figure 11.16) that permits to safely attain the gains needed to detect minimum ionizing particles (Büttner *et al.*, 1998); up to five stages have been tested. With the upper GEM side coated with a photosensitive layer, the device can detect single photoelectrons and is used for imaging of Cherenkov radiation (see Section 14.6).

A set of up to 20 Triple-GEM detectors, 30 × 30 cm^2 of active area each, has been installed in the COMPASS spectrometer at CERN and operates since 2001; the detectors have a fast charge readout and recording on perpendicular strips at 400 μm pitch, and provide localization

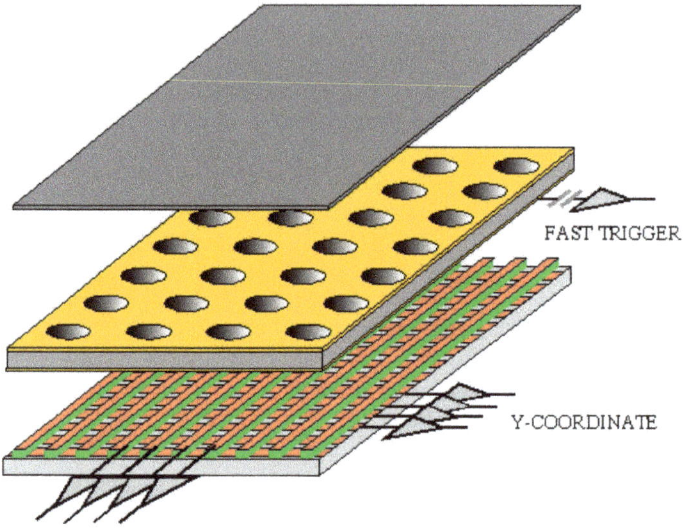

Figure 11.15. Single-GEM detector.

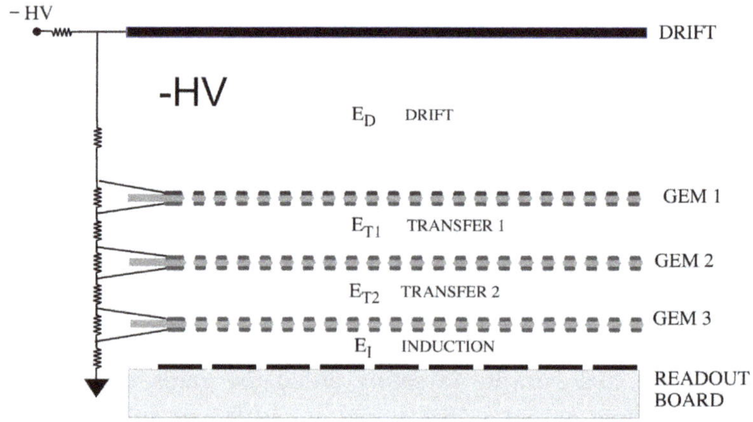

Figure 11.16. Schematics of a Triple-GEM detector.

accuracies for fast charged particles around 70 μm. Figure 11.17 is a view of the detectors installed in the beam (Altunbas *et al.*, 2002).

The GEM foils manufacturing process permits, using custom masks, to realize electrode shapes fitting the experimental needs. Semi-circular detectors have been designed for the TOTEM experiment, a small angle tracker in the forward arms of the CMS experiment, Figure 11.18. Around

Figure 11.17. COMPASS Triple-GEM chambers (Picture CERN).

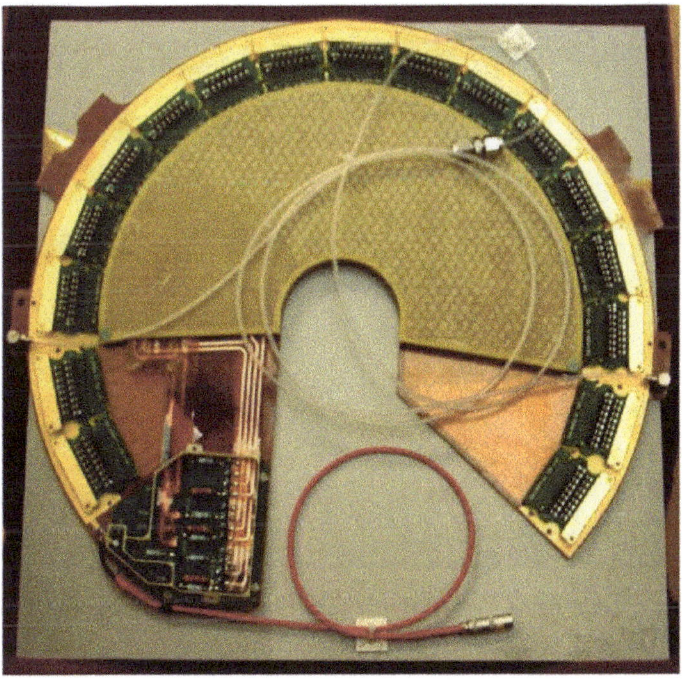

Figure 11.18. A Triple-GEM module for TOTEM (Picture CERN).

thirty cm in diameter, the triple-GEM modules have on the anode plane 256 concentric strips providing the radial coordinate and ~1,500 readout pads used for triggering. Two sets of eight identical modules provide radial and azimuthal coordinates of charged particle tracks created in the interaction vertex (Figure 11.19).

To follow the current luminosity increase of CERN's LHC collider, GEM detectors arrays are replacing rate-limited systems based on wire chambers, namely for the CMS and ALICE experiments. Figure 11.20 shows GEM electrodes of various sizes being manufactured at CERN to instrument the CMS forward muon detector, and Figure 11.21 the installation of a Triple-GEM module on the muon endcap (Abbaneo *et al.*, 2019). The large number of devices has required the development of a new assembly method for the chambers and the outsourcing of the production to several collaborating institutes (Bianco, 2020).

Large volume gaseous detectors, as time projection chambers, are seriously affected by the increase in particle flux due to the accumulation

Figure 11.19. One half of the TOTEM GEM tracker before installation; an identical module is mounted from the left side (Picture CERN).

Figure 11.20. A view of CERN's GEM production line for the CMS endcap muon detector upgrade (Picture CERN).

Figure 11.21. Installation of a GEM module on the CMS forward muon detector (Picture CERN).

Figure 11.22. Installation of a GEM module replacing the MWPC readout in the ALICE TPC (Picture CERN).

of positive ions in the drift space. The ions backflow can be reduced to a few percent using as amplifying element a stack of GEM foils with offset holes; this is the solution adopted for the upgrade of the ALICE TPC, where the MWPC-based readout was replaced with multi-GEM chambers. To minimize the dead space caused by the chamber's frames, the new modules were mounted from inside out, as shown in Figure 11.22. The upgrade was completed in 2020 (Adolfsson *et al.*, 2021).

The flexibility of the GEM electrodes permits the manufacturing of non-planar devices: since charge multiplication occurs in the holes, bending of the foil does not affect the main detector properties. Built and operated by the KLEO-2 experiment at the INFN laboratory in Frascati (Italy), the four-layers cylindrical GEM tracker shown in Figure 11.23 provides sub-mm azimuthal and longitudinal spatial resolutions using anodic readout strips at 25° in respect to the axis (Balla *et al.*, 2013).

While developed originally for the needs of particle physics, thanks to their superior high-rate and position accuracy performances GEM-based detectors have found many applications in other fields: plasma diagnostic, medical imaging, astrophysics; for a review see (Sauli, 2016).

Figure 11.23. The cylindrical GEM detector for the KLOE-2 experiment (Balla *et al.*, 2013).

The photolithographic process used to manufacture the GEM foils enables the realization of holes diameters and pitch of hundred microns and below, needed for achieving high position accuracies. A coarser manufacturing method using mechanical or laser drilling on thicker supports, albeit providing reduced resolutions, has the advantage of being self-supporting and more robust. Named Optimized GEM (Periale *et al.*, 2002), Thick GEM (Chechik *et al.*, 2006) or Large Electron Multiplier (LEM) (Badertscher *et al.*, 2008), the electrodes have been used in applications where rigidity and robustness are paramount requirements.

11.5 Hybrid Multi-Pattern Gaseous Detectors

Notwithstanding their high-rate capability and robustness, MPGD-based devices have limited multi-track resolution if using projective strips readout. Anodes with a matrix of pads have been employed, but are restricted in size by the number of electronic channels required. Aiming at the development of an instrument to measure the polarization of X-rays in astrophysics applications, Ronaldo Bellazzini and collaborators at Pisa University coupled a fine-pitch GEM amplifying structure to a dense matrix of narrow readout pads, Figure 11.24 (Bellazzini *et al.*, 2003). A custom large scale integration ASIC chip with hexagonal readout pads at 80 μm pitch is used to collect and record the charge, permitting the reconstruction of short photoelectron tracks and to determine their creation point and emission angle, that depends on the photon polarization, Figure 11.25 (Bellazzini *et al.*, 2004).

A fully equipped detector is mounted on NASA's Imaging X-ray Polarimetry Explorer (IXPE) satellite, launched in 2021 has already permitted to image the X-ray emission from extragalactic sources (Baldini *et al.*, 2021).

A similar device named GridPix or Ingid was developed by Harry van der Graaf and collaborators at the National Institute for Subatomic Physics (NIKHEF) in Amsterdam, coupling a gaseous amplification device to a high-density pixel readout integrated chip developed for solid-state silicon sensors (van der Graaf, 2007). As shown in Figure 11.26 the

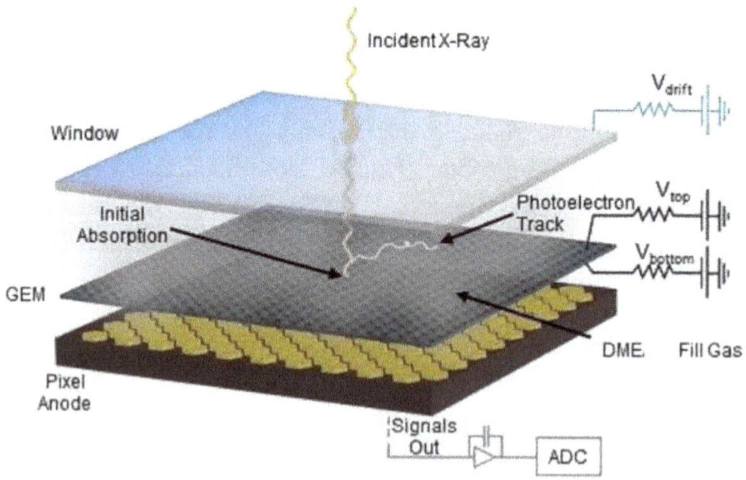

Figure 11.24. X-ray polarimeter. Ionized trails released by the conversion of a soft X-ray are amplified by a GEM electrode and revealed by a fine pad structure (Bellazzini *et al.*, 2003).

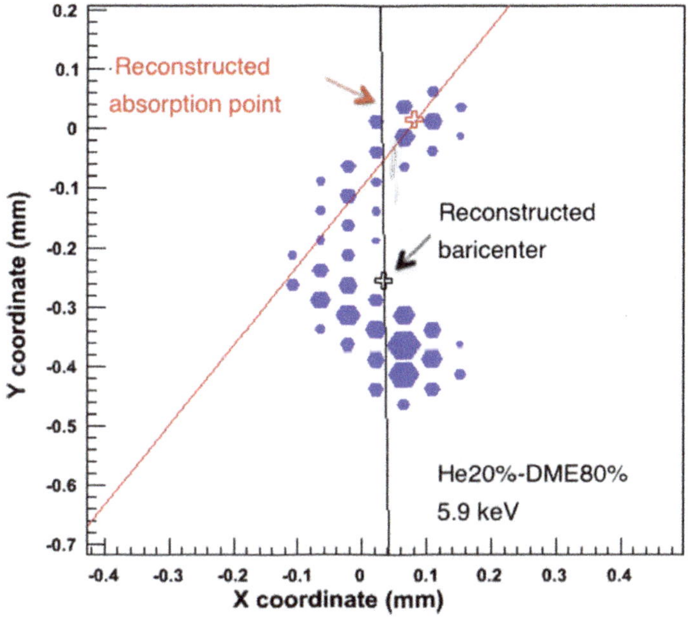

Figure 11.25. Image of a 5.9 KeV photoelectron; the weight of the pixels is proportional to the recorded charge (Bellazzini *et al.*, 2013).

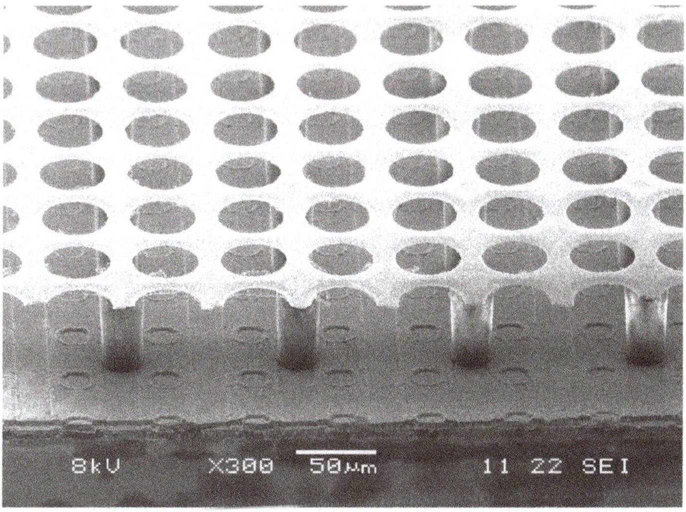

Figure 11.26. GridPix hybrid MPGD, coupling a MICROMEGAS-like multiplier to a TIMEPIX pixel sensor (Koppert *et al.*, 2013).

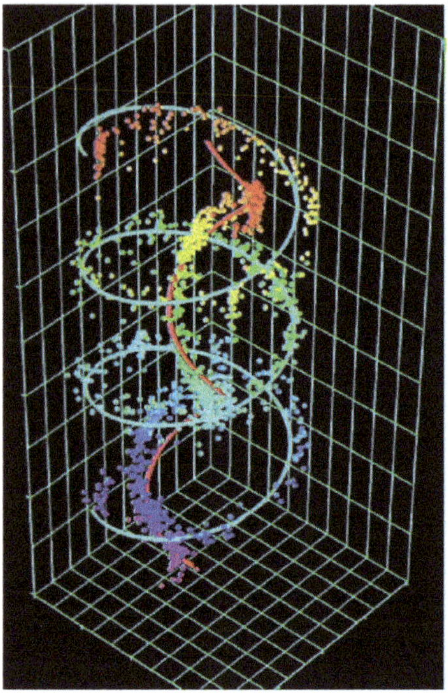

Figure 11.27. Image of electrons spiralizing in a magnetic field, recorded with the InGrid detector. The pixel size is 55 × 55 μm (van Der Graaf, 2010).

detector is built with wafer post-processing to construct a metallic grid over a TIMEPIX silicon chip, supported by insulating pillars and effectively creating a MICROMEGAS structure. Ionization electrons released in the upper gas layer drift into the holes, multiply and are collected, amplified and recorded by the chip readout structure (Koppert *et al.*, 2013). The low noise of the silicon pixels and the high charge gain of the gaseous amplifier permit the detection of single electrons; Figure 11.27 illustrates the imaging capability of the GridPix detector recording ionized trails from an electron source spiralizing in magnetic field (van Der Graaf, 2010).

Aiming at the readout of large TPC chambers, a pixel detector has been developed based on an assembly of four TIMEPIX-3 chips with 512 × 512 pixels at 55 μm pitch (Ligtenberg *et al.*, 2020).

Chapter 12

Optical Imaging Chambers

With the exception of the scintillating proportional counters, described in Chapter 7, most gaseous counters exploit charge multiplication and collection to detect ionization yields, demanding the use of large amounts of electronics channels to record the events. Developed in the late eighties by Charpak and collaborators, an alternative method exploits the light emitted by some gases in the near visible range, imaged using commercially available solid-state cameras directly or with the use of suitable wavelength shifters. Figure 12.1 is an example of beam tracks recorded with an imaging TPC-like chamber filled with a gas mixture containing triethylamine (TEA), a vapour emitting scintillation photons around 260 nm (Suzuki *et al.*, 1988).

The introduction in the late nineties of micro-pattern gaseous chambers, and the verification of the large scintillation yields of carbon tetrafluoride at visible wavelengths has led to the development of a new class of detectors named imaging chambers (Pansky *et al.*, 1995). Albeit restrained in rate by the limits of recording cameras, imaging chambers have found numerous applications in fields where image quality and low cost are paramount issues (Fraga *et al.*, 2001). Figure 12.2 is an image of thermal neutron interacting in an ^3He-CF_4 recorded with a GEM optical detector (Fraga *et al.*, 2002); other examples are energy-resolved fluorescence analysis (Veloso *et al.*, 2010), X-ray tomography and beam monitoring (Brunbauer *et al.*, 2018). Figure 12.3 shows a hadron shower

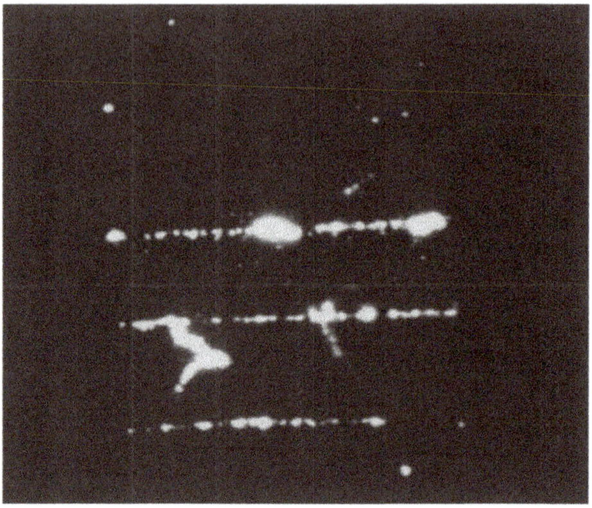

Figure 12.1. Beam tracks recorded with an optical imaging TPC (Suzuki *et al.*, 1988).

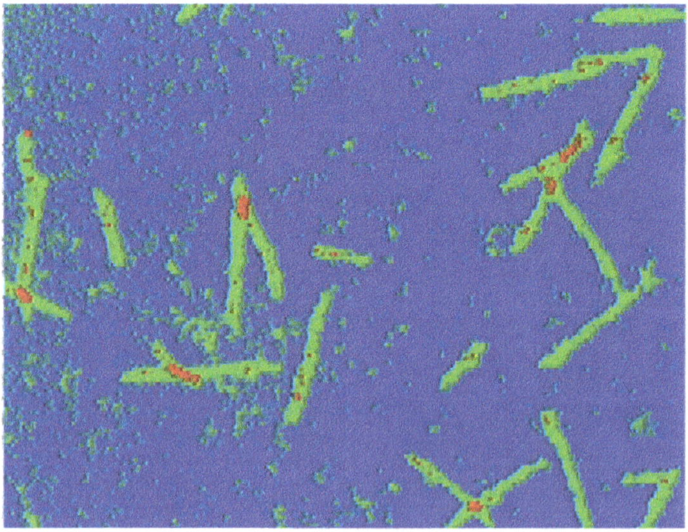

Figure 12.2. Neutron interaction in ^{3}He recorded with an optical imaging detector. Several proton-triton events are visible (Fraga *et al.*, 2002).

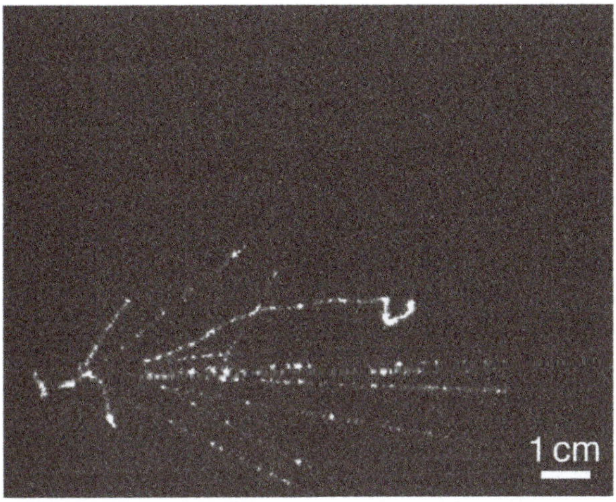

Figure 12.3. Hadronic shower with a GEM imaging detector (Brunbauer *et al.*, 2018).

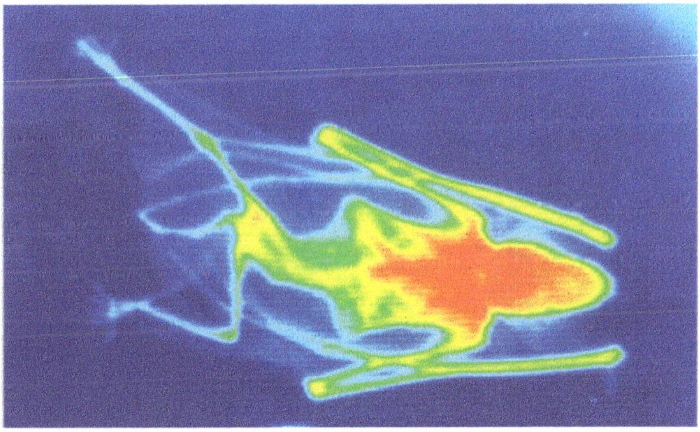

Figure 12.4. Soft X-ray radiography of a small bat recorded with a glass MICROMEGAS imaging detector (Brunbauer *et al.*, 2020).

recorded with an optical imaging chamber consisting in a Triple-GEM operated with an argon-CF_4 mixture and read-out with a solid state CCD camera; Figure 12.4 is a soft X-ray absorption radiography of a small mammal, recorded with a transparent anode glass MICROMEGAS (Brunbauer *et al.*, 2020).

Chapter 13

Cryogenic and Dual-Phase Detectors

Gaseous counters, efficient in the detection and localization of even small energy losses by radiation, are inadequate for the detection of events having very small interaction probability with matter such as neutrinos, hypothetical weakly interacting massive particles (WIMPS), dark matter and the like. Increasing the counter volume and pressure helps but encounters practical limits. With their higher densities, liquid noble gases permit to largely increase the interaction probability. For a review of science and technology of liquid argon detectors see (Cavanna *et al.*, 2018) and (Bonivento and Terranova, 2024).

Proposed in 1977 by Nobel laureate Carlo Rubbia, the liquid argon time projection chamber relies on the observation that in highly purified LAr electrons released by ionizing tracks can be transported without losses over long distances under the effect of an electric field. The process of detection exploits the signal induction by the electrons crossing successive wire plane at angles, the time of the event being provided by photomultipliers detecting the fast argon scintillation.

The largest of its kind, the ICARUS T600 LAr TPC, after extensive testing the National Institute of Nuclear Physics (INFN) at the University of Pavia, was installed and operated from 2010 in the underground Gran Sasso Laboratory. Consisting of two independent volumes sharing a common cathode and with about 500 tons of active mass, it permitted the detection of neutrinos generated by CERN's accelerator in Geneva contributing to the search for $\nu_\mu \rightarrow \nu_e$ oscillations (Rubbia *et al.*, 2011).

Figure 13.1 shows the ICARUS T600 installation, and Figure 13.2 the first reconstructed neutrino interaction, recorded in 2010.

Figure 13.1. The ICARUS T600 complex installed in the underground Gran Sallo laboratory (Antonello *et al.*, 2015).

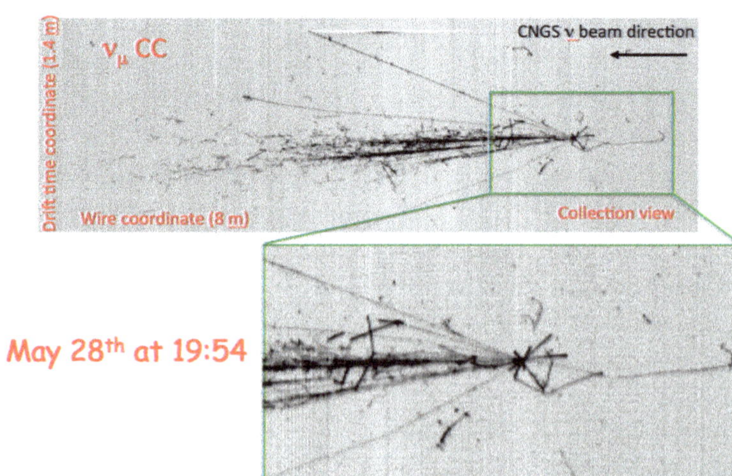

Figure 13.2. First interaction of a neutrino generated by the CERN to Gran Sasso long baseline neutrino beam in 2010 (Antonello *et al.*, 2015).

Moved in 2014 to CERN, it was refurbished before being sent to Fermi National Laboratory (FNAL) in Batavia to operate on the short baseline neutrino beam (Antonello *et al.*, 2015). Concurrent devices are in construction to study neutrino oscillations, the 100 tons Micro Booster Neutrino Experiment (MicroBooNE) (Acciarri *et al.*, 2017) and the Deep Underground Neutrino Experiment (DUNE) with a Near Detector at FNAL and a 10,000 tons LAr Far Detector located at Sanford Underground Research Facility (SURF) in Lead, South Dakota (Abi *et al.*, 2020).

In LAr TPCs the electron charge is merely collected; for large detection volumes, this requires extreme levels of purity to avoid capture losses and very low noise amplifiers. Developed in 2008 at the Institute for Particle Physics (ETH Zurich), the dual-phase TPCs overcomes these limitations (Badertscher *et al.*, 2008). With a suitable choice of electrodes and fields, electrons released and drifting in the liquid are extracted into an overlying gas phase where they can be amplified by charge multiplication. Figure 13.3 shows the electrodes and field configuration in the dual-phase TPC; the liquid is separated from the gaseous phase by a pair of

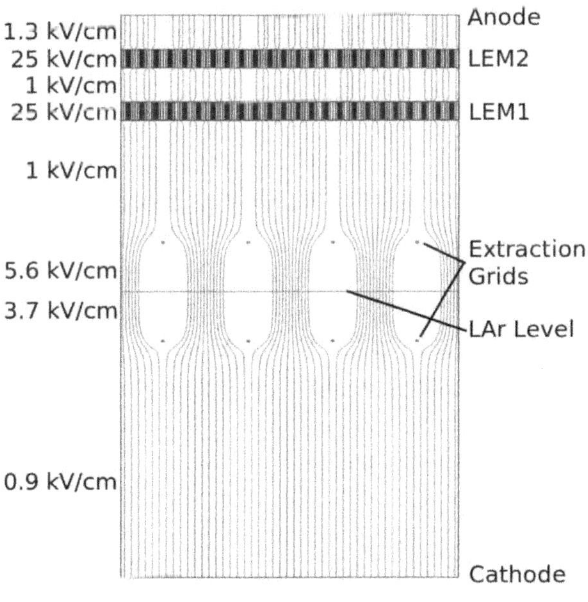

Figure 13.3. Electric field configuration in the dual-phase LAr TPC (Badertscher *et al.*, 2008).

Figure 13.4. The dual-phase LAr TPC before insertion in the cryostat (Badertscher *et al.*, 2011).

wire grids at appropriate potentials, ensuring extraction of the electrons, further drifted into a cascade of large electron multipliers (LEM) for charge multiplication and detection at the anode. The picture in Figure 13.4 shows the first prototype of a dual-phase TPC, with 21 cm drift length. This design provides 3D-tracking and energy loss capable of adjustable charge amplification, a promising readout technology for next generation neutrino detectors and direct Dark Matter searches.

The dual-phase concept has been extended to Liquid Xenon TPCs (Aprile *et al.*, 2004) and the detection of electroluminescence emitted by the high field in the gas phase (Agnes *et al.*, 2018). For an exhaustive overview of dual-phase TPCs for imaging rare processes see (González-Díaz, Monrabal and Murphy, 2018). A dual-phase option of the 10 kTons LAr TPC for the DUNE is under development (Chardonnet, 2020).

Chapter 14

Particle Identification

14.1 Methods

Most of the detectors described in previous chapters permit to reveal the particles' spatial positions. To perform identification, one or more kinematic quantities have to be measured, such as energy loss, momentum and velocity. Figure 14.1 indicates the approximate regions covered by various identification methods as a function of the charged particle momentum. The exploitable process depends on the particle's energy and mass:

- **Magnetic analysis**: charged particles bending in magnetic field, providing momentum P and charge Q;
- **Differential energy loss (*dE*/*dx*)**: charged particles, low to medium energy;
- **Calorimetry**: measurement of the charged or neutral particle total energy E;
- **Threshold Cherenkov counters**: charged particles with velocity above a threshold value $\beta > \beta_T$;
- **Cherenkov Ring Imaging (RICH)**: charged particles above threshold, provides the velocity β;
- **Time-of-Flight**: particle velocity β;
- **Transition radiation**: photons emission of relativistic particles crossing the interface between media of different dielectric constant; allows to discriminate between electrons and hadrons.

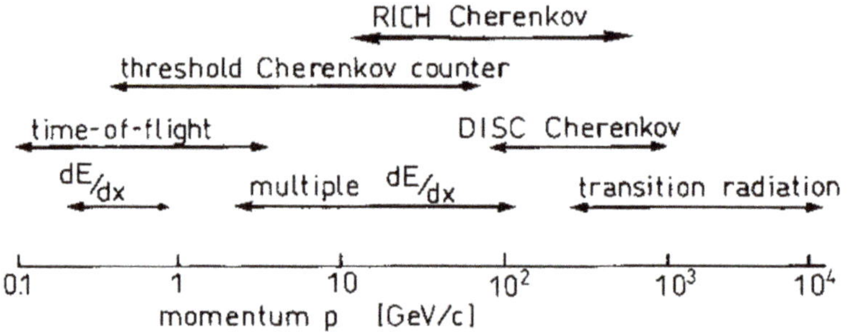

Figure 14.1. Approximate regions of particle identification from various processes.

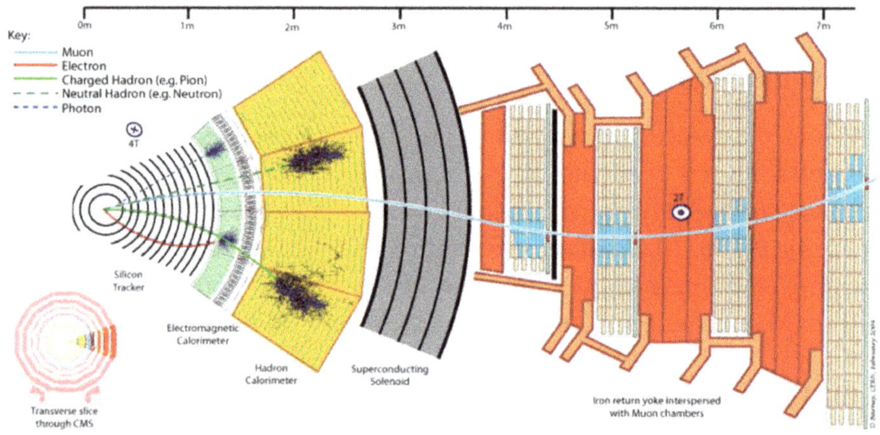

Figure 14.2. Schematic cross section of one sector of CMS with indication of the various detectors used for particle identification.

In high energy physics experiments, various methods are combined to provide wide range particle identification, as shown schematically in Figure 14.2 for one sector in the Compact Muon Solenoid (CMS) experiment at CERN. For a detailed description of the identification methods and corresponding detectors see for example (Lippmann, 2012).

14.2 Magnetic Analysis

Under the effect of external magnetic fields, the motion of charged particles deviates from the original trajectory due to the Lorentz effect, a force component perpendicular to the plane containing the particle velocity and the magnetic field vector. A measurement of the deflection provides the transverse momentum of the charged particle in the plane of curvature; the sign of the deflection gives the charge. For a unitary charge particle and a bending of radius ρ in a homogeneous magnetic field B, the momentum is $P = mv = B\rho$. Expressing momentum and field in practical units, the expression becomes: P(GeV/c) = $B.\rho$ (Tesla.m). Images of tracks bending were given in previous chapters.

The error in the momentum determination depends on the length and precision of the measurement of the particle's trajectory. Particle identification is then achieved by the simultaneous measurement of the velocity-dependent differential energy loss (see the next section).

14.3 Differential Energy Loss

Charged particles release energy through a series of Coulomb interactions with the molecules, creating a trail of ion-electron clusters. Occasionally, the electrons created in the primary encounters further ionise the medium, leading to a total ionisation count exceeding two or three times the number of primaries; often, the ejected electron has enough energy to create a visible secondary track, named delta electron.

The specific energy loss per unit track length, also named linear stopping power or differential energy loss (dE/dx), is a function of the particle velocity and depends on the composition and density of the medium. A classic expression developed in the 1930s of last century by Hans Bethe and perfected by Felix Bloch provides the average energy loss in materials as a function of particle velocity; it has a characteristic shape, decreasing with the particle velocity and reaching a minimum before a small increase, named relativistic rise. Expressing the material thickness in reduced units (length by density) the specific energy loss as a function of energy depends on the particle mass but little from the medium, as shown in Figure 14.3. Figure 14.4 provides the relative energy loss in the relativistic rise region, normalised to the minimum, as a function of momentum for several particles in Ar-CO_2, often used in gaseous detectors.

A detailed description of the interaction processes between radiation and matter is outside the scope of this book. The Review of Particle Physics (Workman *et al.*, 2022) has a chapter dedicated to the outcomes

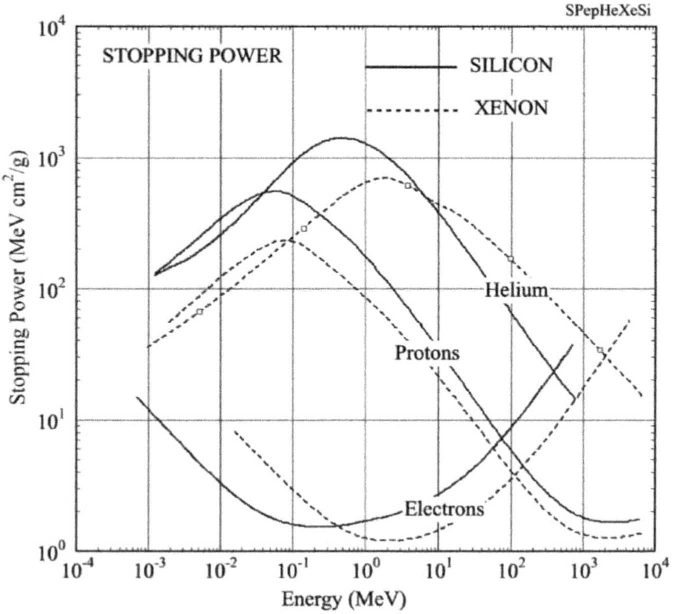

Figure 14.3. Specific energy loss, or stopping power, of electrons, protons and alpha particles in silicon and xenon as a function of the particle energy (Berger *et al.*, 2011).

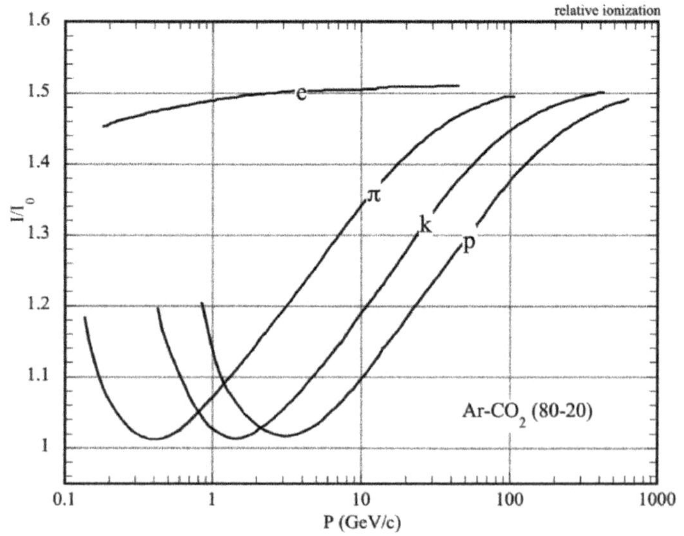

Figure 14.4. Relative energy loss of protons, kaons, pions, and electrons, normalized to the minimum.

of the passage of particles through matter. Web-based calculators permit to compute range and energy loss in materials of protons, electrons and helium ions (Berger *et al.*, 2011), photons (Hubbell and Seltzer, 1996) and neutrons (*Neutron Cross Section of the elements*, 2017).

The total number of ion-electron pairs released per unit length can be estimated from the ratio between energy loss and a material-dependent average ionization energy, W_I ; approximate values of W_I for fast, singly charged particles in several materials are given in Table 14.1 (from different sources).

The Bethe-Block expression provides the mean values of energy loss; the statistics of the interactions, with formation of clusters and sporadic emission of long-range delta electrons, results in a spread around the average. For fast particles, the energy loss distribution has a characteristic shape described by the Landau expression, with a long tail at values above the average.

Examples of distributions for fast particles are shown in Figure 14.5 for a thin silicon slab and for gaseous detectors; the resolution is better in the solid-state device, due to the larger ionization yield.

As discussed in the section on large volume drift chambers, time projection chambers in particular (Section 9.3) the resolution of the energy loss determination is improved by a statistical analysis over many

Table 14.1. Average ionisation energy for relativistic particles in materials. For gases, the table provides primary and total ion pairs at normal conditions (20°C, 1 bar).

GAS	W_I *(eV)*	$N_P(cm^{-1})$	$N_T(cm^{-1})$
H_2	37	5.2	9.2
He	41.3	3.5	8
Ne	30	13	50
Ar	25	25	106
Xe	22	41	312
CH_4	30	37	54
C_2H_6	26	48	112
Semiconductor	*W(eV)*		
Si	3.62		
Ge	2.96		
GaAs	4.2		

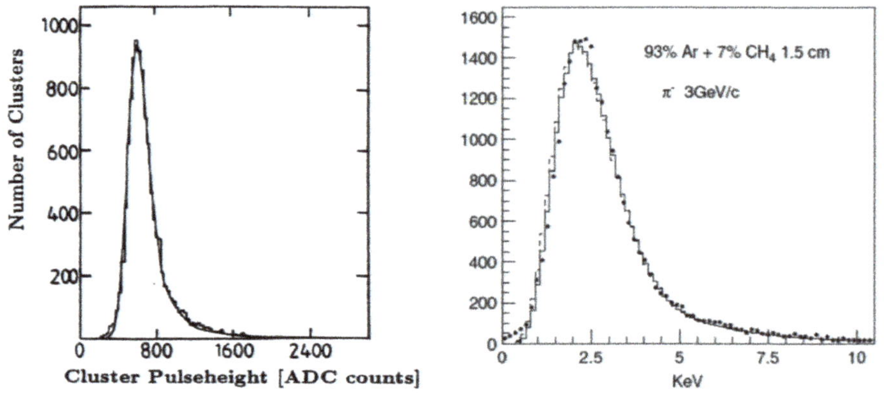

Figure 14.5. Left: ionisation energy loss of fast particles in a 280 μm thick silicon detector (Caccia *et al.*, 1987); right: in 1.5 cm Ar-CH_4 at NTP (20°C, one bar) (Smirnov, 2005).

individual samples recorded for each track. Coupled to the measurement of momentum from the track's bending in magnetic field, a plot of the average energy loss as a function of momentum permits identification of the particle mass in the region up to a few GeV/c, Figure 14.6.

Charged particles traverse matter until losing all their energy; the depth of penetration, or range, can be estimated with the sources quoted

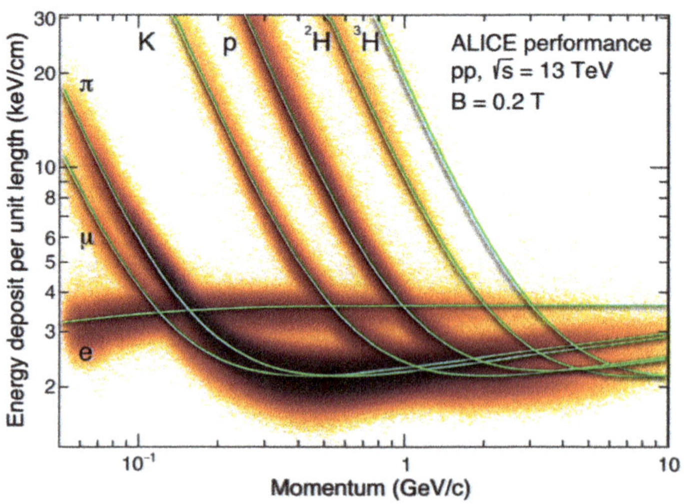

Figure 14.6. Differential energy loss as a function of momentum measured in the ALICE TPC (Hauer, 2022).

above. The differential energy loss has an increase, named Bragg peak, at the end of the range; Figure 14.7 is an example of computed stopping power at the end of range for 70 MeV protons in water (Bichsel, 1995). Figure 14.8 shows the differential ionisation loss of MeV α particles as a function of penetration in gases at pressures slightly above atmospheric

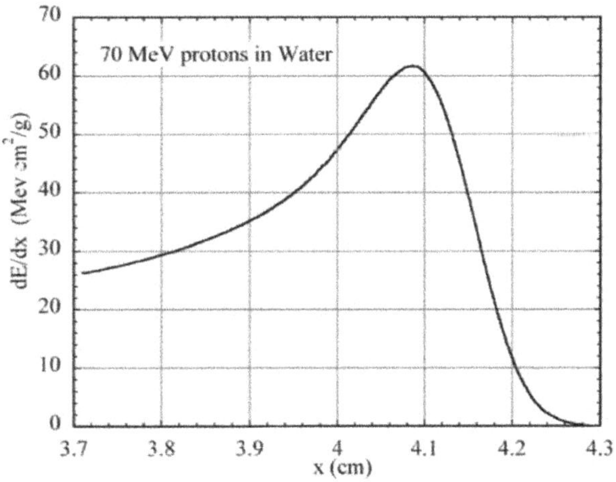

Figure 14.7. Differential energy loss of protons at the end of range in water (Bichsel, 1995).

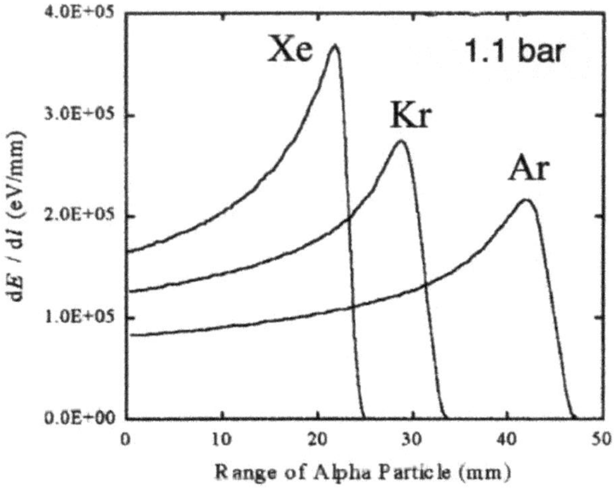

Figure 14.8. Differential energy loss in gases of MeV α particles at the end of range (Saito *et al.*, 2002).

(Saito *et al.*, 2002). Alpha-emitting radioactive sources are often used for testing gaseous detectors.

The large increase of energy loss in the Bragg peak is exploited in deep tumour hadron therapy; for a review see (Amaldi and Braccini, 2011).

14.4 Calorimetry

The term "calorimetry" denotes the measurement of the total energy of a particle, charged or neutral, released in a suitable absorbing medium. After the first interaction, a cascade of secondary processes gradually develops a shower of particles, charged and neutral, gradually dissipating the initial energy by multiple ionizations and excitations of the medium. A measurement of the overall ionization and/or scintillation yields provides the total energy of the projectile. Photons and leptons (electrons and positrons) have only electromagnetic interactions with molecules, while hadrons experience both electromagnetic and nuclear interactions. In this case, a fraction of the energy loss, spent in the creation of neutrons and various nuclear processes, may go undetected resulting in an intrinsically worse resolution than for photons and leptons. As shown by the simulation of Figure 14.9 the transverse extension of a hadronic shower is wider, causing losses for limited size calorimeters.

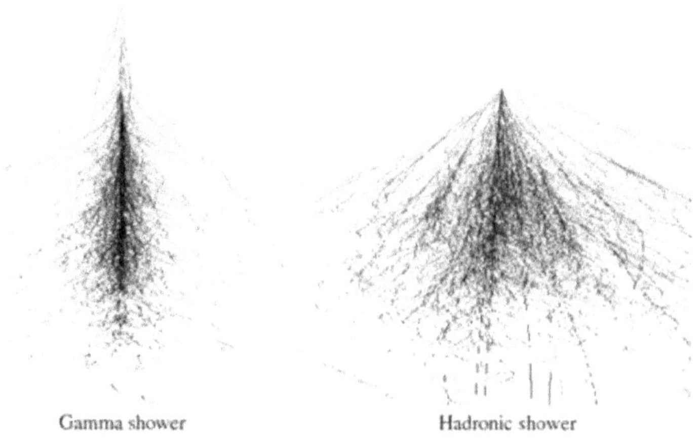

Figure 14.9. Simulated longitudinal and transverse development of an electromagnetic and hadronic shower in an uniform absorber Figure 14.9 (Cherkashyna *et al.*, 2015).

The physics and practice of electromagnetic and hadronic calorimeters are discussed in the reviews by Chris Fabjan and Fabiola Gianotti[1] (Fabjan and Gianotti, 2003), David Barney (Barney, 2020) and the 2017 book by Richard Wigmans (2017).

Particle calorimeters belong to two main categories: homogeneous, where the energy is dissipated in a uniform detector, and sampling, where the active detectors are sandwiched between heavy conversion slabs. Examples of the first class are sodium iodide and other crystals where the energy loss is revealed from scintillation using optical detectors (see Chapter 5). Expensive but providing the best energy and time resolution in the detection of hard photons and gammas, they are used as electromagnetic calorimeters in high energy particle physics. An example is the composite electromagnetic calorimeter of CMS at CERN shown schematically in Figure 14.10, consisting in a central barrel and two endcap modules instrumented with lead tungstate crystals with avalanche photodiodes or vacuum phototriodes readout.

Crystal-based detectors are used in demanding medical applications such as computer assisted (CAT) and positron emission tomography (PET).

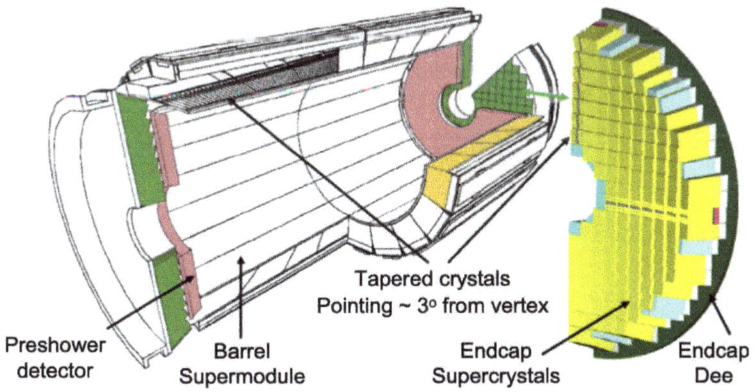

Figure 14.10. The CMS barrel and end-cap electromagnetic calorimeter, comprising 76,000 lead tungstate crystals covering close to 4π solid angle (Brown, 2007).

---

[1] Former project leader and spokesperson of the CMS experiment, in 2016 Gianotti was nominated CERN's Director General, a mandate she will hold until 2025.

Less performing but more affordable for large detection volumes, sampling calorimeters consist in a succession of high-Z converter plates interleaved with active modules: plastic scintillators, solid state sensors, liquid or gaseous ionization chambers. An example is the combined electromagnetic-hadronic calorimeter of the ATLAS experiment at CERN, with copper converters and liquid argon as ionization detector, Figure 14.11; the converter plates have an "accordion" geometry, designed to minimize cracks and dead gaps, Figure 14.12.

Owing to the losses due to undetected nuclear processes, the energy resolution for hadrons is typically a factor of two worse than for electrons. Using as converter a fissile material like depleted uranium, interleaved with other materials, partly recovers the losses. Simulation and experimental efforts have been invested to find the optimum choice in the development of the so-called compensating calorimeters; the quality of the improvement is expressed as ratio of electrons to hadron response, e/h,

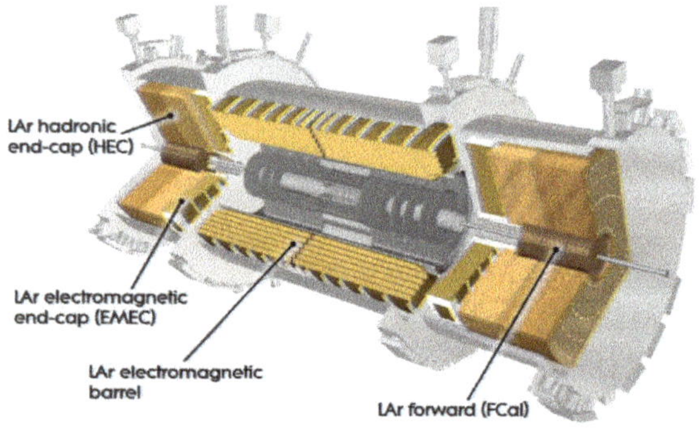

Figure 14.11. The ATLAS liquid argon calorimeter complex (Wilkens, 2009).

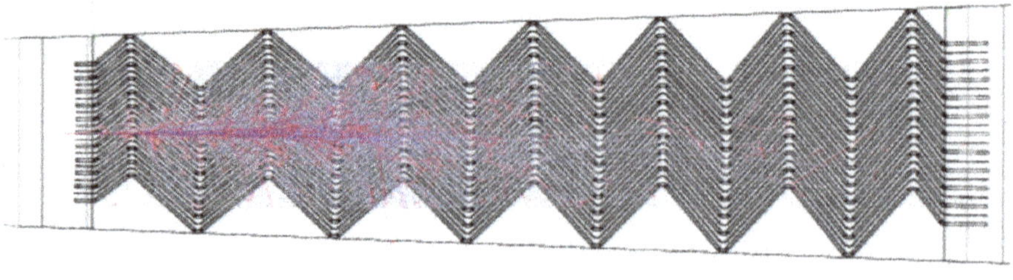

Figure 14.12. The "accordion" geometry of the ATLAS LAr-Cu converters (Wilkens, 2009).

ideally equal to unity. Examples are the fine-sampling uranium-copper scintillator calorimeters deployed at CERN's Intersecting Storage Rings (Akesson *et al.*, 1985) and the UA1 uranium-tetramethyl pentane (TMP) ionization chambers at CERN's proton-antiproton collider (Evans, 1991) achieving an e/h value close to one at energies above 1 GeV. Due to the difficulties and cost of handling the depleted uranium this approach was essentially abandoned.

14.5 Time-of-Flight

A particle with mass m and momentum P has a relativistic velocity:

$$\beta = \frac{P}{\sqrt{P^2 + m^2}}$$

the time to travel over a distance L is then $T = L/c\beta$.

The time interval between detection in two aligned sensors depends on the particle's velocity. For two particles of mass m_1 and m_2 the time-of flight difference over a distance L is:

$$\Delta T = \frac{L}{c}\left(\sqrt{1+\frac{m_1^{\,2}}{P^2}} - \sqrt{1+\frac{m_2^{\,2}}{P^2}}\right)$$

Computed from this expression, Figure 14.13 shows the time-of-flight difference between pairs particles as a function of momentum for one metre distance. Easily achieved below 1 GeV/c, particle identification requires sub-ns time resolutions for higher momenta.

The particle identification power, defined as ratio of the difference between measured quantities and resolution, depends then on the accuracy of the time-of-flight measurement, σ_{TOF}. Figure 14.14 shows the separation power for several pairs of particles, computed for a baseline of 3.5 m and three values of the time resolution (Lippmann, 2012).

The TOF resolutions achievable with various detector technologies are discussed in Refs. (Klempt, 1999) and (Williams, 2012); only a few examples will be give here. Some devices, while very effective, have only been demonstrated at small-scale sizes; others have been scaled-up to the size needed to equip large systems.

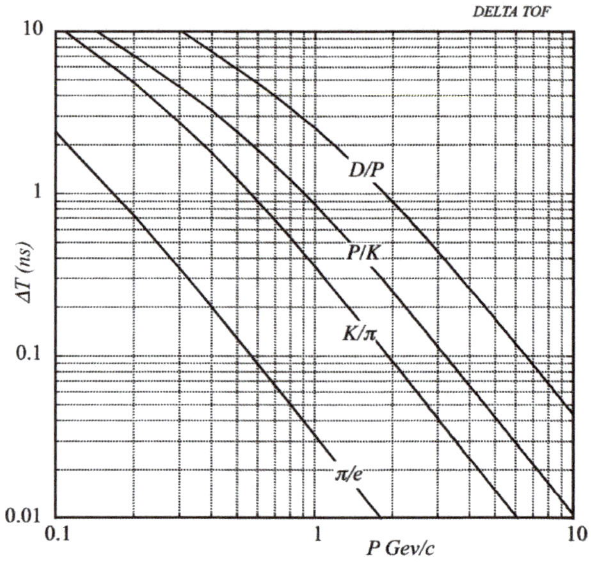

Figure 14.13. Time-of-flight difference between pairs of particles as a function of momentum for 1 m distance.

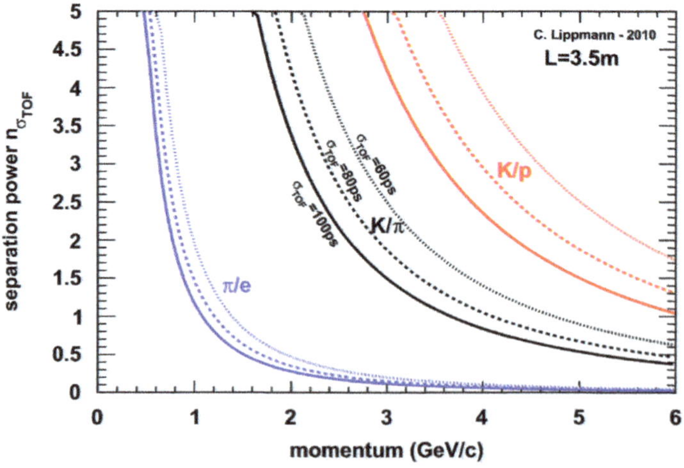

Figure 14.14. Separation power for pairs of particles as a function of momentum (Lippmann, 2012).

The time resolution of a radiation detector is a convolution of dispersive factors: the statistics of the energy loss process, the variations in response of the sensors, the noise of the electronics. Gaseous detectors suffer from the fluctuations in collection time of the ionization yields losses, that is reduced for denser solid state devices; scintillation counters are affected by the prompt fluorescence and decay times. The best performance has been be obtained exploiting the Cherenkov emission in a radiator, occurring for charged particles above a threshold velocity.

- A resolution of 6 ps has been obtained in a 3 GeV/c pion beam with a small size detector exploiting the Cherenkov light from a quartz radiator and a Micro-Channel Plate Photomultiplier (MCP) (Inami *et al.*, 2006).
- Fast inorganic scintillators, read-out with photomultipliers, MCPs or Silicon Photomultipliers. Using LYSO crystals readout with SiPM, the AXIAL PET achieved a time resolution of around 130 ps, adequate for medical applications (Casella *et al.*, 2015).
- With a small size PICOSEC detectors, coupling a Cherenkov radiator with a photosensitive MICROMEGAS gaseous multiplier, time resolutions down to 25 ps have been demonstrated in extensive beam tests (Bortfeldt *et al.*, 2018). The measurements are been extended with improved 100-channels, 100 cm^2 devices (Utrobicic *et al.*, 2023).
- Single-gap gaseous Resistive Parallel Plate Counters (RPC) can reach time resolutions better than 50 ps, albeit at reduced efficiency; multi-gap counters (MRPC) ensure full detection efficiency for fast particles, with equal or better time resolution (Fonte, Smirnitski and Williams, 2000). This is the solution adopted in large experiments: ALICE at CERN ((Akindinov *et al.*, 2012), STAR at RHIC (Llope, 2005) and others (see Chapter 10).

14.6 Cherenkov Threshold and Ring Imaging Counters

Charged particles traversing a medium of refraction index n emit a coherent forward front of photons if their velocity exceeds the velocity of light in the medium, $\beta = v/c \geq 1/n$. The effect was observed in 1934 by Pavel

Table 14.2. Refraction index for a range of materials.

Material	n
Glass	1.4
Aerogel	1.022
Neopentane	1.00177
Argon (1 bar)	1.000135

Cherenkov, at Lebedev Physical Institute in Moscow; in 1958 Cherenkov shared the Nobel Prize in Physics with Ilya Frank and Igor Tamm for the discovery.

For a velocity β above threshold, photons are emitted at an angle θ given by $cos\ \theta = 1/n\beta$; the maximum emission angle is $\theta_C = arcos\ (1/n)$. Values of the refraction index for a range of radiators used in particle physics are given in Table 14.2. Photons are emitted in the range of wavelengths for which the radiator is transparent; their number is given by $N = N_0\ L\ sin^2\ \theta$, where N_0 is a figure of merit depending on the radiator, typically 50 to 100 cm^{-1} and L the radiator length.

Figure 14.15 shows the computed Cherenkov angle as a function of velocity in a range of values of the refraction index of the medium (Gilmore, 1980).

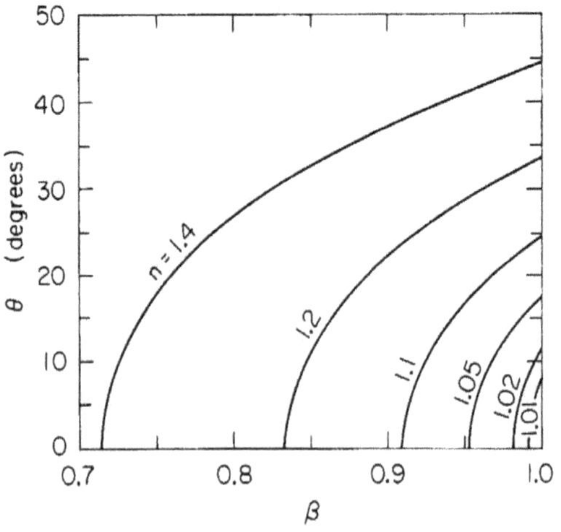

Figure 14.15. Photons emission angle as a function of the particle velocity for a range of values of the refraction index of the radiator.

For a particle perpendicular to the interface between two media, to permit detection the Cherenkov angle should be smaller that the critical angle for total internal reflection, given by $\theta_C = arcsin(n_2/n_1)$ where n_1 and n_2 are the refraction indices of the two materials.

Cherenkov counters are universally used in physics to identify the mass of charged particles in beams. In the simplest configuration, they consist in a gas-tight vessel where the radiation is emitted and reflected by a mirror to a photomultiplier (Figure 14.16 left). An improved version of the threshold counter focusing the light on collimating slits to select a range of angles and named Differential Isochronous Self-Collimating (DISC) (Figure 14.16 right) permits selection of velocity over a pre-defined range and has been operated both with gaseous and liquids radiators (Meunier *et al.*, 1962). The index of refraction of the medium depends on the gas, and can be changed according to needs varying its pressure; Figure 14.17 shows a gas Cherenkov counter before installation in a particle beam at CERN. Recorded on a non-separated negative particle beam, Figure 14.18 is an example of counting rates of signals over threshold as a function of gas pressure. Several counters in-line at different pressures can be used for full particle recognition. The review article by Litt and Meunier is a comprehensive summary of the subject (Litt and Meunier, 1973).

Named DIRC (Detection of Internally Reflected Cherenkov Light), a new type of ring-imaging Cherenkov detector has been developed for hadronic particle identification in the BABAR experiment at the Stanford Linear Accelerator Center (SLAC) B Factory (PEP-II) (Adam *et al.*, 2005); the geometry of a counter is shown in Figure 14.19. Figure 14.20 is a view of the complete assembly. Each radiator is a long, thin fused silica bar with rectangular cross section; a track with velocity β passing through the radiator with refractive index n emits Cherenkov radiation in

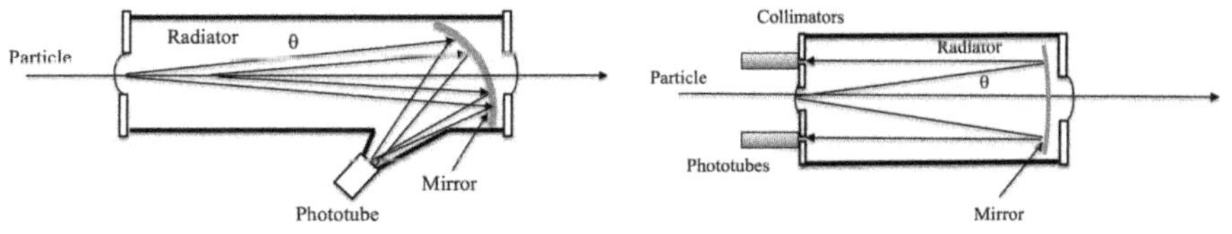

Figure 14.16. Threshold and differential Cherenkov counters.

Figure 14.17. A gas Cherenkov counter (Picture CERN).

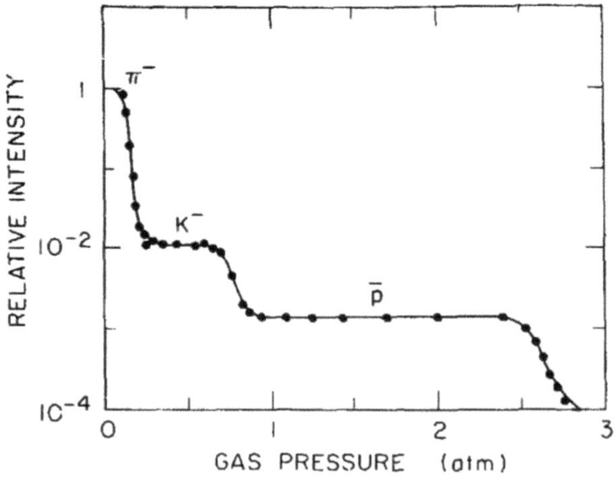

Figure 14.18. Counting rate of a threshold Cherenkov counter as a function of gas pressure in a non-separated positive charged particle beam.

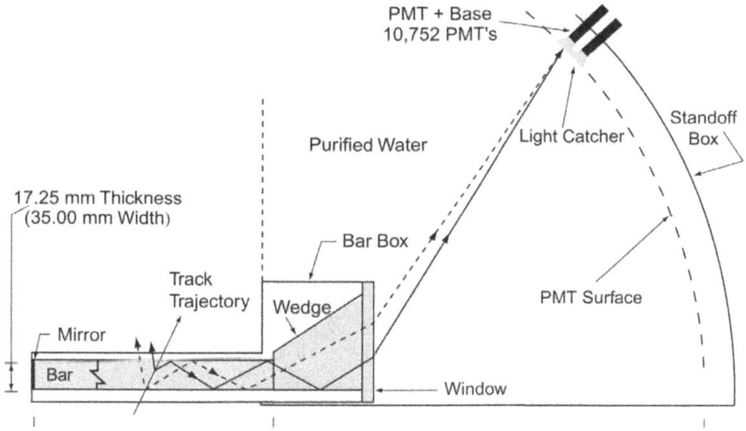

Figure 14.19. The DIRC particle identifier (Adam *et al.*, 2005).

a cone with opening angle θ_C around the particle trajectory. The angle, position, momentum and timing of the track are provided by tracking devices located in front of the radiator, and by the crossing time of the colliding beams in the machine. The photons are detected by an array of densely packed photomultiplier tubes; the position of detection provides the Cherenkov angle in respect of the particle trajectory (Ratcliff and Va'vra, 2020).

The LHCb RICH upgrade at CERN, making use of a matrix of multi-anode photomultipliers as photon detectors, is described in Section 5.3.

Developed in the late seventies, photosensitive multiwire proportional chambers permit to detect and localize single photons. Introduced in the late seventies by Thomas Ypsilantis, the Ring Imaging Cherenkov Counter (RICH) provides the image of the radiation on a circular pattern whose radius depends on the particle velocity (Ypsilantis, 1981). As shown in Figure 14.21 the counter's geometry can be reflective, making use of a spherical mirror to form the ring pattern on the detector image plane, or proximity focusing exploiting the distance from radiator and detector. The ring pattern in the second case is more diffuse, reducing the resolution of the velocity determination.

The reflective geometry is particularly appropriate for use as barrel detector for colliders, and has been adopted by the first large RICH detector of the DELPHI experiment at CERN (Arnold *et al.*, 1988).

Figure 14.20. The DIRC detector at SLAC (Courtesy Jerry Va'vra).

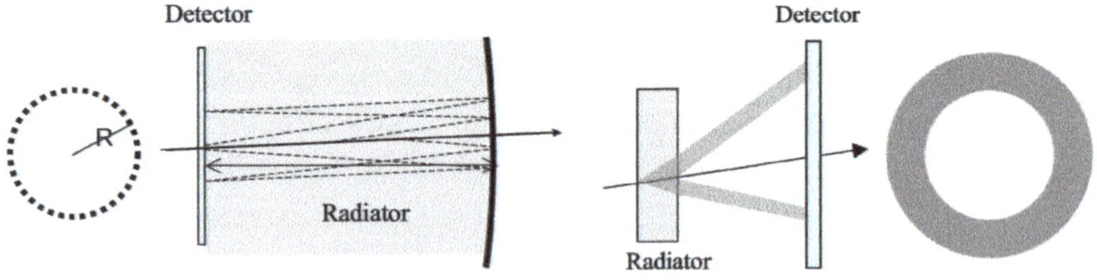

Figure 14.21. Reflective and proximity focusing RICH counters.

A large research effort has been devoted finding gases and vapours suitable for the imaging of photons emitted by Cherenkov effect; the region of sensitivity is limited by the transparency of the window separating the radiator from the detector. The first generation of RICH counters used triethyl amine (TEA[2]), with a photoionization threshold of 7 eV and calcium fluoride windows (Charpak *et al.*, 1979); the discovery of TMAE,[3] a vapour with 5.5 eV threshold, permitted the use of more affordable quartz windows (Anderson, 1980).

A problem encountered using photosensitive vapours in gaseous counters is the emission by the avalanches of VUV photons, reconverting in the gas and inducing secondary effects and discharges. The multistep chamber, Figure 14.22 sharing the amplification process in two cascaded stages, permits to reach the gains needed to detect single electrons, (Charpak and Sauli, 1978). With a few percent addition of TEA to the gas mixtures, Cherenkov photons transmitted by the window convert in the first drift space, are preamplified in a high-field gap and proceed to the second stage of amplification, a MWPC with projective coordinate readout. Owing to the few hundred microns mean absorption length of TEA, photons generated by the avalanche process are absorbed before reaching the drift gap and experience full amplification. Figure 14.23 shows one of the multi-step proportional chambers used for particle identification in the experiment E605 at Fermilab (McCarthy *et al.*, 1986). Mounted on a 15-m long helium-filled radiator with two ports, the detector had a composite CaF_2 window with 32 crystals assembled to cover an active area of 40×80 cm^2. Projective readout on anode and cathode wires at angles permits the unambiguous reconstruction of multi-photon events and

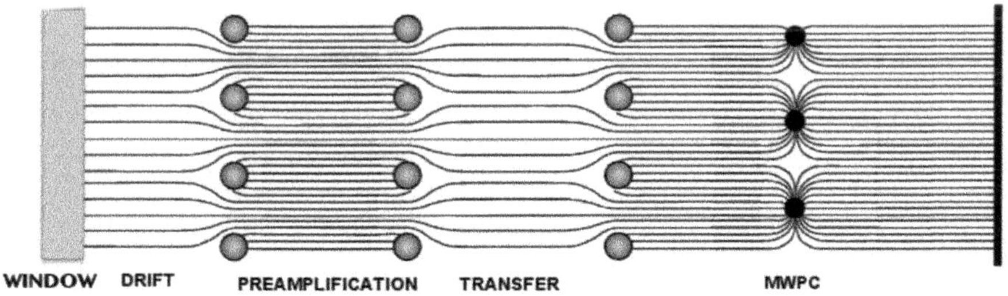

Figure 14.22. Field lines in the multistep chamber. Photoelectrons created in the drift gap are transported into a high field region where they multiply; a fraction of the charge transfers to the MWPC where it is further amplified and detected.

---

[2]Triethyl amine, $(C_2H_5)_3N$.

[3]Tetrakis-dimethyl amino ethylene $[(CH_3)_2 N_2]C$.

Figure 14.23. One of the multistep proportional chambers of the RICH particle identifier of experiment E605 at Fermilab (Picture by the author, 1980).

determination of the radius of the Cherenkov rings, Figure 14.24; Figure 14.25 shows an example of the particle identification power of the detector. The spectrometer has been used to study high-transverse-momentum charged hadron production in 400-GeV/c and 800-GeV/c proton-proton and proton-deuteron collisions.

One of the four LEP experiments, DELPHI, was equipped with ring imaging Cherenkov (RICH) detectors providing μ/K/p identification from 0.17 to 35 GeV/c over nearly 4π solid angle. To cover this momentum range requires a proximity focusing liquid and a reflective gas Cherenkov radiator, as shown schematically in Figure 14.26; fused quartz windows are used to separate the radiators from the drift camber detector containing the photosensitive vapour TMAE. Photoelectrons

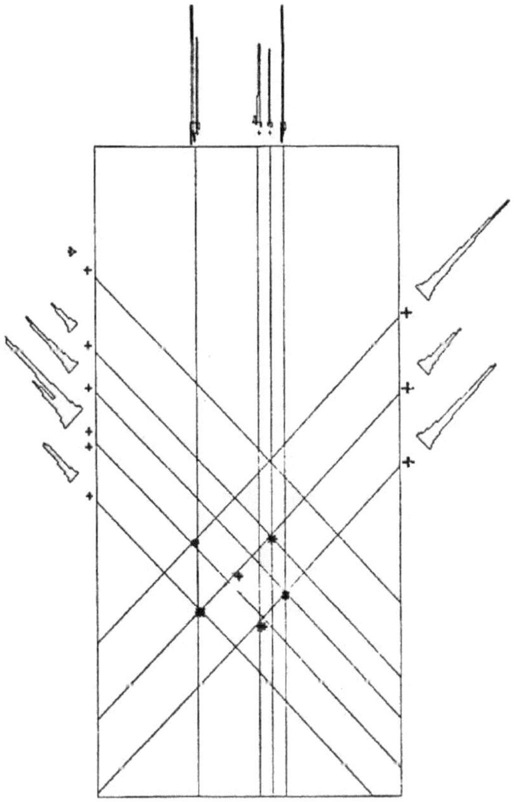

Figure 14.24. A five-photons Cherenkov ring recorded with the E605 imager (McCarthy *et al.*, 1986).

created in the sensitive gap drift to an end-cap multiwire chamber structure where they are amplified and detected (Arnold *et al.*, 1988). With a conveniently low photoionization threshold, 5.4 eV, permitting to use quartz windows, TMAE has however the drawback of a low vapour pressure, 0.87 Torr at 35°C, corresponding to an absorption length of 14 mm. Differing to the case of TEA, photons emitted by the avalanches have a long path in the gas and can induce secondary counts and discharges. The problem has been solved encasing the amplifying wires in cathode mini-tubes with a narrow slit to collect the photoelectrons, while absorbing most of the secondary photons, Figure 14.27. Figure 14.28 shows

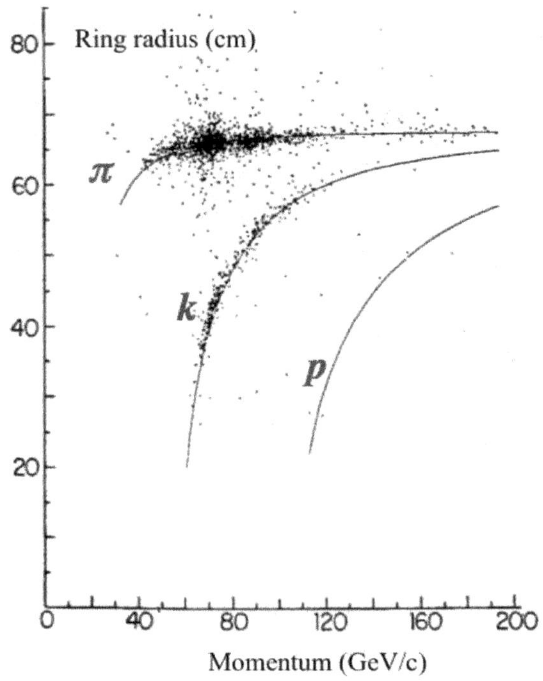

Figure 14.25. Reconstructed Cherenkov ring radius as a function of particle momentum (Adams *et al.*, 1983).

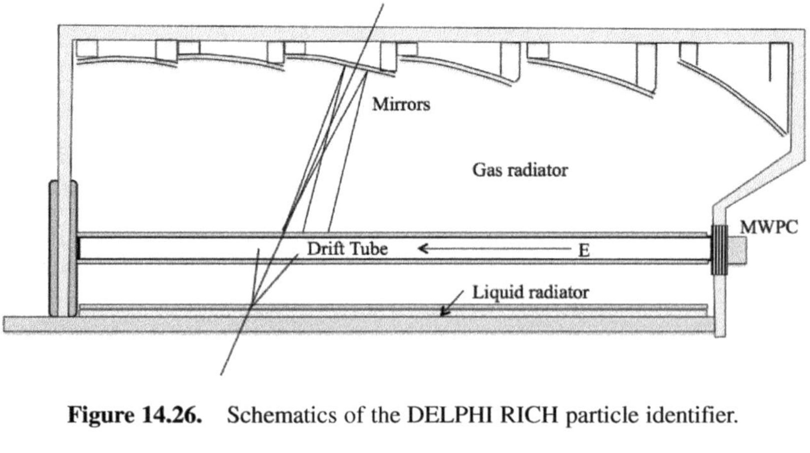

Figure 14.26. Schematics of the DELPHI RICH particle identifier.

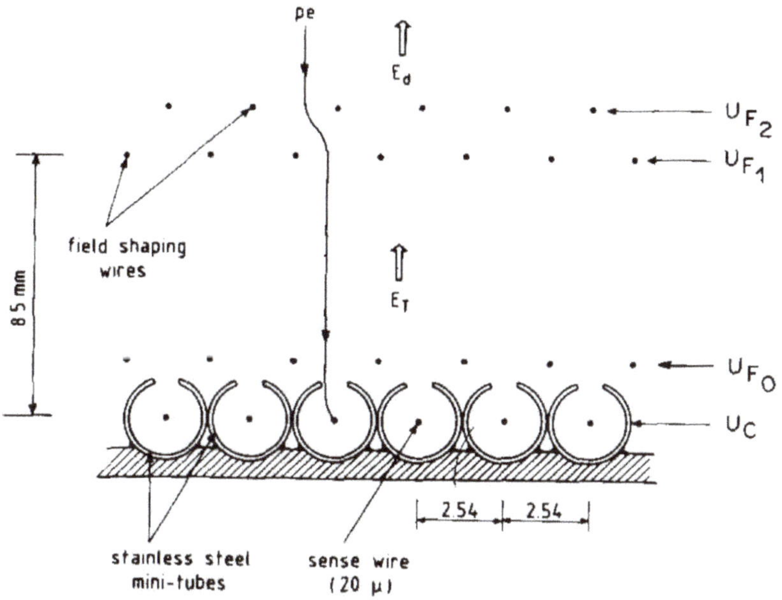

Figure 14.27. Encapsulated wires of the DELPHI RICH detector (Arnold *et al.*, 1988).

Figure 14.28. One sector of the DELPHI RICH photon detector (Picture CERN).

Figure 14.29. The completed DELPHI RICH detector before installation in the experiment (Picture CERN).

one module of the detector, and Figure 14.29 the complete system before installation at CERN. Figure 14.30 gives an example of the particle identification capability of the detector.

A particle identifier of similar conception, named CRID, has been in operation for many years at SLAC (Va'vra, 1999).

Efficient in detecting VUV photons, photosensitive vapours pose a considerable challenge in their use due to chemical reactivity with the detector's components; TMAE, moreover, reacts with oxygen and water, forming harmful deposits.[4] A better solution was found with the use of

[4]TMAE was originally developed as tracer for missiles, owing to its property of fluorescing at visible wavelengths in air.

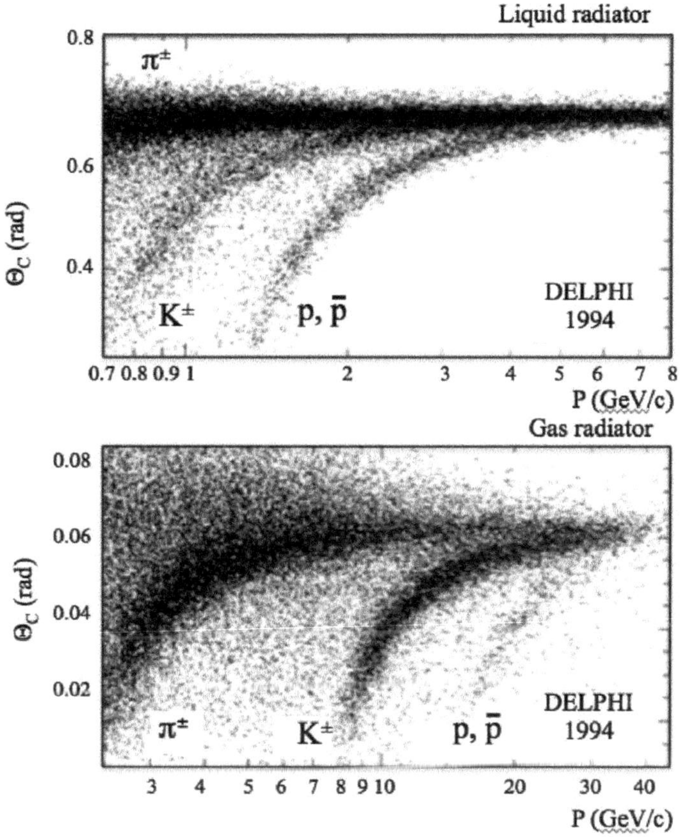

Figure 14.30. Reconstructed Cherenkov angle as a function of particles momentum (Abreu *et al.*, 1996).

solid materials, such as cesium iodide, sensitive to UV photons above 5.7 eV and reaching a quantum efficiency of 40% at 7 eV, near the transparency cut-off of fused silica; for a summary of the progress in manufacturing methods of thin-layer CsI photocathode see for example (Schyns, 2002).

The basic design of a MWPC with a reflective CsI photocathode suitable for proximity focusing RICH is shown in Figure 14.31, as adopted by several experiments: ALICE and COMPASS at CERN, HADES at GSI Darmstadt, STAR at BNL and others. Figure 14.32 is the structure of the COMPASS RICH-1 particle identifier at CERN: a 3 m long Cherenkov

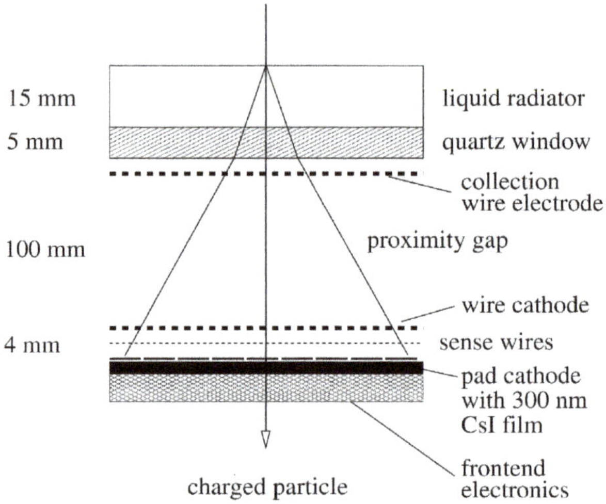

Figure 14.31. MWPC with reflective CsI photocathode (Cozza *et al.*, 2003).

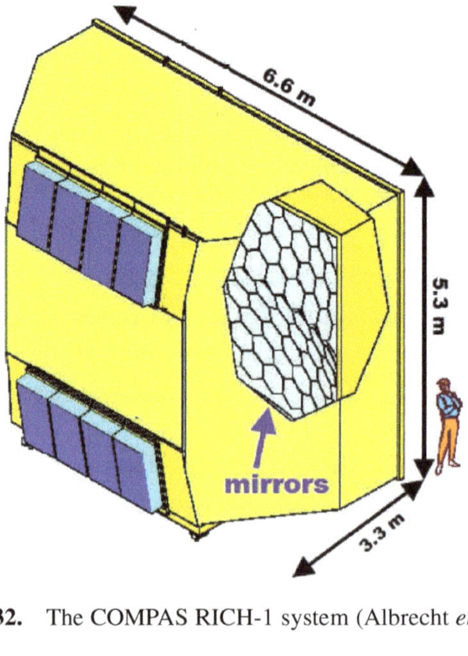

Figure 14.32. The COMPAS RICH-1 system (Albrecht *et al.*, 2003).

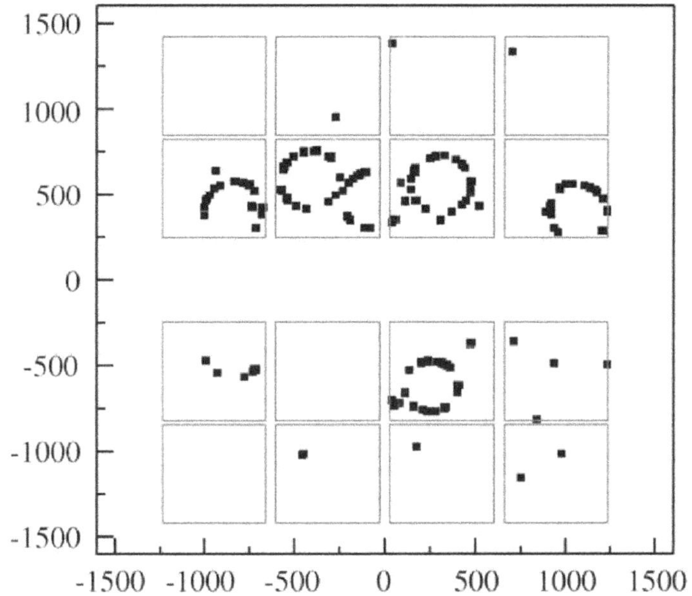

Figure 14.33. Cherenkov photons rings recorded with COMPASS RICH-1 (Albrecht *et al.*, 2003).

radiator vessel filled with C_4F_{10} at atmospheric pressure, a matrix of 116 spherical mirrors reflecting the ring images on a set of eight CsI-coated MWPCs.[5] The detector started operation in 2002, providing π-K separation from 3 to 55 GeV/c over ±200 mrad angular acceptance, at high rates. Figure 14.33 is an example of multiple photon rings recorded with the detector (Albrecht *et al.*, 2003).

The observed long-term detector instabilities and performance degradation, mostly due to discharges and to the reactivity of positive ions with the CsI photosensitive layer, led to the evolution of a more performing RICH-1 upgrade, installed and operated from 2016. The detector scheme shown in Figure 14.34, employing the new generation of micro-pattern gaseous chambers (see Chapter 11), includes two cascaded Thick-GEMs and a MICROMEGAS as final amplifier; the CsI layer is deposited on the top surface of the first TGEM. This geometry permits

[5] It should be noted that glass or crystals cannot be used for perpendicular tracks due to the total internal photon reflection at the radiator-air interface.

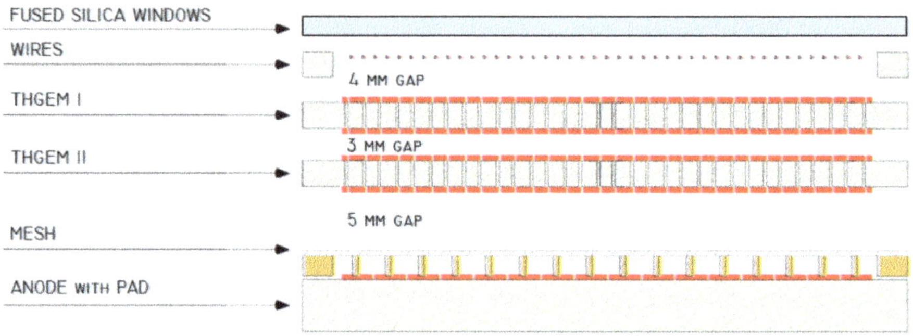

Figure 14.34. Hybrid MPGD detector for the COMPASS-RICH1 upgrade (Alexeev *et al.*, 2014).

reaching the gains needed to detect single photoelectrons, reducing the positive ions feedback hitting the photosensitive layer to a few percent (Alexeev *et al.*, 2014).

Exploiting the different threshold for Cherenkov photons emission, the Hadron Blind Detector (HBD) built and operated by the PHENIX experiment at BNL permits to select electrons from pions. As seen in Figure 14.35 the detector consists in a semi-cylindrical radiator vessel with sets of CsI-coated Triple-GEM modules on the outer radius; the

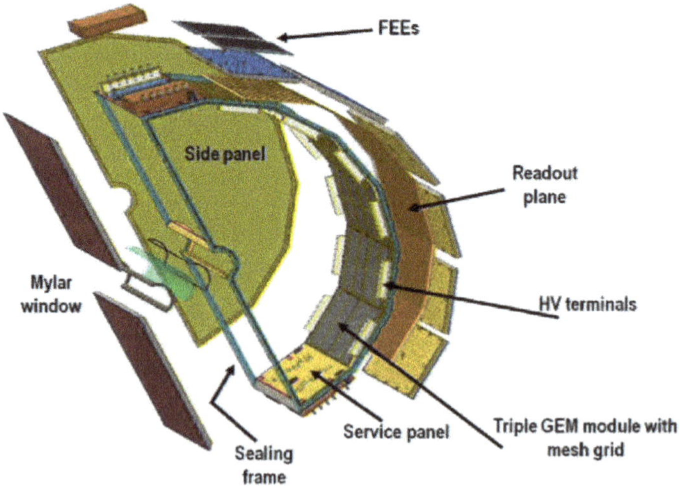

Figure 14.35. The Hadron Blind Detector (Anderson *et al.*, 2011).

electric field applied to the sensitive volume is reversed, as compared to a conventional drift chamber, pushing the electrons away from the readout plane. With a pure CF_4 gas filling at atmospheric pressure, the detector effectively allowed to identify electrons from hadrons in the same momentum range. Commissioned in 2009, it permitted to study electron pairs production in Au-Au central collisions at 200 GeV.

14.7 Transition Radiation

Demonstrated theoretically in 1945 by Vitali Lazarevich Ginsburg and Ilya Mikhailovich Frank, Transition Radiation (TR) occurs when a charged particle crosses the boundary between two media having different dielectric constants. For highly relativistic particles ($\gamma \gtrsim 1000$), the emitted radiation is peaked forward relative to the particle direction and extends into the X-ray domain, where the radiator is transparent, and can be detected; many interfaces are needed to increase the yield. For exhaustive reviews of TR properties and detectors see for example (Dolgoshein, 1993, Andronic and Wessels, 2012). Figure 14.36 is an example of X-ray radiation spectra recorded in a mixed pions-electron beam with a xenon-filled proportional counter, with and without a radiator made with a stack of plastic foils; pions and electrons are well identified.

The TR yield has been studied for a wide choice of multiple regular and unstructured radiators. Since many interfaces are necessary to produce a suitable yield, the materials need to have a low X-ray absorption

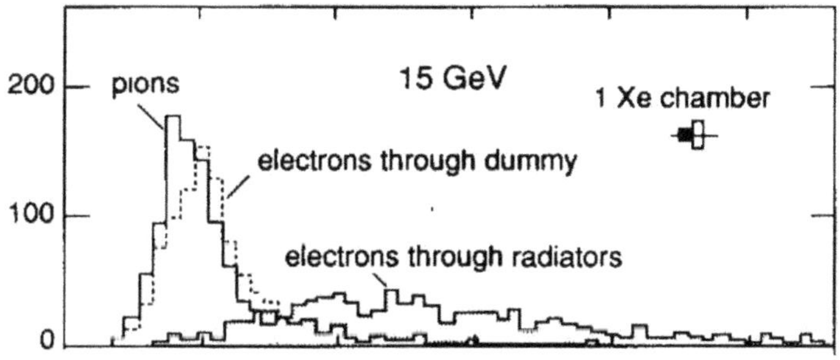

Figure 14.36. X-rays spectra recorded with a Xe-filled counter with and without a TR radiator (Dolgoshein, 1993).

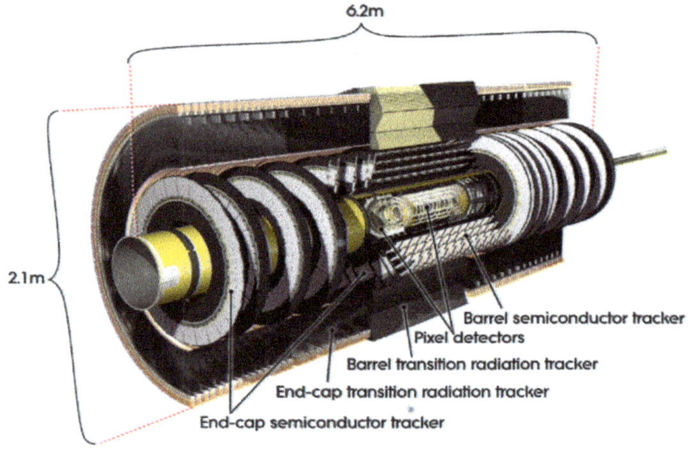

Figure 14.37. ATLAS inner detector at CERN's LHC with the barrel and end-cap TR trackers (Mindur, 2017).

coefficient; Li, Be, polypropylene and mylar foils have been used, as well as expanded plastic foams or fibres; examples are given in Ref. (Brigida *et al.*, 2005). The detector itself can be a gaseous counter with a Xe filling.

The ATLAS Transition Radiation Tracker at LHC is the outermost inner subsystem in the 2T solenoid field. It consists of arrays of around 30,000 straws made with multilayer Kapton tubes and with a Xe-CO_2–O_2 gas filling, Figure 14.37. Commissioned in 2012, it contributes significantly to the precision of the momentum measurement of charged particles and to the identification of electrons.

The ALICE TRD, covering the full azimuth and the pseudo-rapidity range from –0.84 to 0.84 is part of the ALICE central barrel. It consists of 522 chambers arranged in six layers, Figure 14.38; each chamber comprises a foam/fibre radiator followed by a Xe-CO_2-filled MWPC preceded by a drift region. The radiator modules are made with polypropylene fibre mats sandwiched between two plates of Rohacell,[6] mechanically reinforced by lamination of carbon fibre sheets. The detector started data taking at the first LHC collisions in 2009.

[6] Closed-cell polymethacrylimide hard foam.

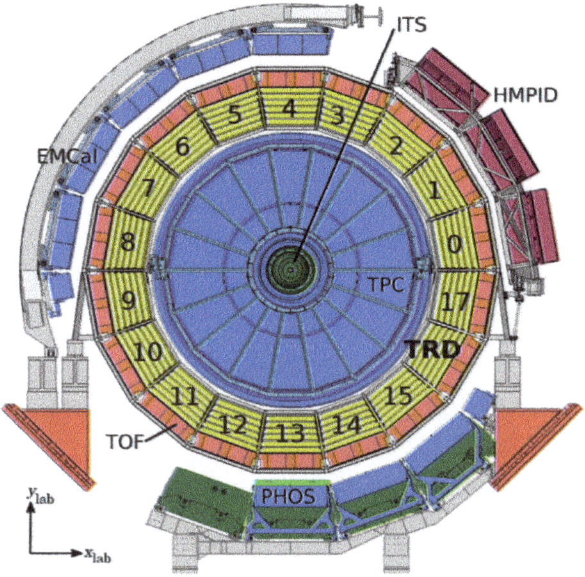

Figure 14.38. Cross section of the ALICE detector with the Transition Radiation Tracker (Acharya *et al.*, 2018).

A TD detector will instrument the Compressed Baryonic Matter (CBM) experiment at the Facility for Antiproton and Ion Research (FAIR) accelerator at GSI (Darmstadt). TR detectors are also used in astroparticle physics as part of balloon and space-borne experiments to study cosmic rays and high-energy particles.

Chapter 15
Concluding Remarks

The progress in our understanding of the basic properties of matter is intimately linked to the development of innovative radiation detection technologies, from the early photographic emulsions to the gigantic and sophisticated detectors instrumenting particle and astrophysics experiments. A considerable effort has been invested to understand properties and operating characteristics of the evolving generations of devices, aiming at the improvement of their efficiency, positioning and rate capabilities, following the concurrent construction of more and more powerful particle accelerators; in astrophysics, the challenge is to build large and massive systems capable to detect rare events. This book describes this evolution, discussing the technological limitations of existing systems and suggesting directions of research towards improving our ability to explore the complexity of the basic elements and forces that constitute our world.

In view of the vast amount of literature dedicated to the research on radiation detectors, whenever possible reference is given to comprehensive reviews on each major subject; a list of basic textbooks completes the bibliography.

Basic Bibliography

Abbrescia, M., Peskov, V. and Fonte, P. (2018). *Resistive Gaseous Detectors* (Wiley VHC, Weinheim).

Blum, W. and Rolandi, G. (1993). *Particle Detection with Drift Chambers* (Springer-Verlag, Berlin).

Christophorou, L. G. (1971). *Atomic and Molecular Radiation Physics* (Wiley, New York).

Da Via, C., Dalla Betta, G.-F. and Parker, S. (2019). *Radiation Sensors with 3D Electrodes* (Taylor & Francis).

Grupen, C. (1996). *Particle Detectors* (Cambridge University Press, Cambridge).

Hartmann, F. (2009). *Evolution of Silicon Sensor Technology in Particle Physics Second Edition* (Springer Tracts in Modern Physics, Vol. 275).

Herz, R. H. (1969). *The Photographic Action of Ionizing Radiation* (Wiley Interscience, London).

Knoll, G. (2010). *Radiation Detection and Measurements, 4th Edition* (Wiley, New York).

Kolanoski, H. and Wermes, N. (2020). *Particle Detectors, Fundamentals and Applications* (Oxford University Press).

Korff, S. A. (1955). *Electron and Nuclear Counters* (Van Nostrand, New York).

Leroy, C. and Rancoita, P.-G. (2012). *Silicon Solid State Devices and Radiation Detectors* (World Scientific, Singapore).

Loeb, L. B. (1961). *Basic Processes in Gaseous Electronics* (University of California Press, Berkeley).

Lutz, G. (1999). *Semiconductor Radiation Detectors* (Springer-Verlag, Berlin).

McGregor, D. S. and Shultis, J. K. (2021). *Radiation Detectors. Concepts, Methods and Devices* (CRC Press).

Meek, J. and Craggs, J. D. (1953). *Electrical Breakdown of Gases* (Clarendon Press, Oxford).

Nappi, E. and Peskov, V. (2013). *Imaging Gaseous Detectors and their Applications* (Wiley).

Price, W. (1958). *Nuclear Radiation Detection* (McGraw-Hill, New York).

Rice-Evans, P. (1974). *Spark, Streamer, Proportional and Drift Chambers* (Richelieu, London).

Ritson, D. M. (1961). *Techniques of High-Energy Physics* (Interscience, New York).

Sauli, F. (2014). *Gaseous Radiation Detectors* (Cambridge University Press, Cambridge).

Sauli, F. (2020). *Micro-Pattern Gaseous Detectors* (World Scientific, Singapore).

Sauli, F. and Oliveri, E. (2024). *Gaseous Detectors Handbook.* http://fabio.home.cern.ch/fabio/publications.res/handbook.pdf

Sharpe, J. (1964). *Nuclear Radiation Detectors* (Methuen, New York).

Tsoulfanidis, N. and Landsberger, S. (2021). *Measurement & Detection of Radiation* (CRC Press).

Viehhauser, G. and Weidberg, T. (2024). *Detectors in Particle Physics* (CRC Press, London).

Wigmans, R. (2017). *Calorimetry-Energy Measurements in Particle Physics,* 2ndEdition (Oxford University Press).

References

Abbaneo, D. *et al.* (2019). Layout and assembly technique of the GEM chambers for the upgrade of the CMS first muon endcap station. *Nucl. Instrum. Methods Phys. Res. A*, 918, pp. 67–75.

Abbrescia, M. *et al.* (1997). Effect of the linseed oil surface treatment on the performance of resistive plate chambers. *Nucl. Instrum. Methods Phys. Res. A*, 394, pp. 13–20.

Abe, K. *et al.* (1997). The SLD VXD detector and its initial performance. *Nucl. Instrum. Methods Phys. Res. A*, 386, pp. 46–51.

Abgrall, N. *et al.* (2011). Time projection chambers for the T2K near detectors. *Nucl. Instrum. Methods Phys. Res. A*, 637, pp. 25–46.

Abgrall, N. *et al.* (2014). NA61/SHINE facility at the CERN SPS: Beams and detector system. *J. Instrum.*, 9, p. P06005.

Abi, B. *et al.* (2020). Long-baseline neutrino oscillation physics potential of the DUNE experiment: DUNE Collaboration. *Eur. Phys. J. C*, 80, p. 978.

Abrams, G. *et al.* (1989). The Mark II Detector for the SLC. *Nucl. Instrum. Methods Phys. Res. A*, 281, pp. 55–80.

Abreu, P. *et al.* (1996). Performance of the DELPHI detector. *Nucl. Instrum. Methods Phys. Res. A*, 378, pp. 57–100.

Acciarri, R. *et al.* (2017). Design and construction of the MicroBooNE detector. *J. Instrum.*, 12, p. P02017.

Acharya, S. *et al.* (2018). The ALICE transition radiation detector: Construction, operation, and performance. *Nucl. Instrum. Methods Phys. Res. A*, 881, pp. 88–127.

Adam, I. *et al.* (2005). The DIRC particle identification system for the BABAR experiment. *Nucl. Instrum. Methods Phys. Res. A*, 538, pp. 281–357.

Adam, W. *et al.* (2021). The CMS Phase-1 pixel detector upgrade. *J. Instrum.*, 16, p. P02027.

Adams, M. *et al.* (1983). π/K/p Identification with a Large-Aperture Ring-Imaging Cherenkov Counter. *Nucl. Instrum. Methods*, 217, pp. 237–243.

Adolfsson, J. *et al.* (2021). The upgrade of the ALICE TPC with GEMs and continuous readout. *J. Instrum.*, 16, p. P03022.

Agafonova, N. *et al.* (2009). The detection of neutrino interactions in the emulsion/lead target of the OPERA experiment. *J. Instrum.*, 4, p. P06020.

Agnes, P. *et al.* (2018). Electroluminescence pulse shape and electron diffusion in liquid argon measured in a dual-phase TPC. *Nucl. Instrum. Methods Phys. Res. A*, 904, pp. 23–34.

Aielli, G. *et al.* (2012). Highlights from the ARGO-YBJ experiment. *Nucl. Instrum. Methods Phys. Res. A,* 661, pp. 550–555.

Aiello, S. *et al.* (2025). Observation of an ultra-high-energy cosmic neutrino with KM3NeT. *Nature*, 638, pp. 376–382.

Aihara, H. *et al.* (1983). Spatial resolution of the PEP-4 Time Projection Chamber. *IEEE Trans. Nucl. Sci.*, NS-30, pp. 76–81.

Akesson, T. *et al.* (1985). Properties of a Fine-Sampling Uranium-Copper Scintillator Hadron Calorimeter. *Nucl. Instrum. Methods Phys. Res. A,* 241, pp. 17–42.

Akindinov, A. *et al.* (2012). The MRPC-based ALICE Time-Of-Flight detector: Commissioning and first performance. *Nucl. Instrum. Methods Phys. Res. A,* 661, pp. S98–S101.

Albrecht, E. *et al.* (2000). Performance of hybrid photon detector prototypes with 80% active area for the rich counters of LHCB. *Nucl. Instrum. Methods Phys. Res. A,* 442, pp. 164–170.

Albrecht, E. *et al.* (2003). COMPASS RICH-1. *Nucl. Instrum. Methods Phys. Res. A.* 502, pp. 112–116.

Alexeev, M. *et al.* (2014). Status and progress of the novel photon detectors based on THGEM and hybrid MPGD architectures. *Nucl. Instrum. Methods Phys. Res. A,* 766, p. 133.

Alexopoulos, T. *et al.* (2019). Performance studies of resistive-strip bulk micromegas detectors in view of the ATLAS New Small Wheel upgrade. *Nucl. Instrum. Methods Phys. Res. A,* 937, pp. 125–140.

Alexopoulos, T. *et al.* (2020). Construction techniques and performances of a full-size prototype Micromegas chamber for the ATLAS muon spectrometer upgrade. *Nucl. Instrum. Methods Phys. Res. A.* 955, p. 162086.

ALICE_Collaboration. (2024). ALICE upgrades during the LHC long shutdown 2. *J. Instrum.* p. P05062.

Allison, W. *et al.* (1974). The identification of secondary particles by ionisation sampling (ISIS). *Nucl. Instrum. Methods,* 119, pp. 499–507.

Allison, W. W. W. W. *et al.* (1984). Relativistic charged particle identification with ISIS2. *Nucl. Instrum. Methods Phys. Res.*, 224, pp. 396–407.

Alme, J. *et al.* (2010). The ALICE TPC, a large 3-dimensional tracking device with fast readout for ultra-high multiplicity events. *Nucl. Instrum. Methods Phys. Res. A,* 622, pp. 316–367.

Altunbas, C. *et al.* (2002). Construction, test and commissioning of the triple-gem tracking detector for COMPASS. *Nucl. Instrum. Methods Phys. Res. A,* 490, pp. 177–203.

Amaldi, U. and Braccini, S. (2011). Present challenges in hadrontherapy techniques. *Eur. Phys. J. Plus,* 126(7), pp. 1–15.

Anderhub, H. *et al.* (2003). Experience with the L3 vertex drift chamber at LEP. *Nucl. Instrum. Methods Phys. Res.*, A515, p. 31.

Anderson, D. F. (1980). A xenon gas scintillation proportional counter coupled to a photoionization detector. *Nucl. Instrum. Methods,* 178, pp. 125–130.

Anderson, M. *et al.* (2003). The STAR time projection chamber: A unique tool for studying high multiplicity events at RHIC. *Nucl. Instrum. Methods,* A499, p. 659.

Anderson, W. *et al.* (2011). Design, construction, operation and performance of a Hadron Blind Detector for the PHENIX experiment. *Nucl. Instrum. Methods Phys. Res. A,* 646, pp. 35–58.

Andronic, A. and Wessels, J.P. (2012). Transition radiation detectors. *Nucl. Instrum. Methods Phys. Res. A,* 666, pp. 130–147.

Antonello, M. *et al.* (2015). Operation and performance of the ICARUS T600 cryogenic plant at Gran Sasso underground Laboratory. *J. Instrum.* 10, p. P12004.

Antoni, T. *et al.* (2004). A large area limited streamer tube detector for the air shower experiment KASCADE-Grande. *Nucl. Instrum. Methods Phys. Res. A,* 533, pp. 387–403.

Anulli, F. *et al.* (2002). The BaBar instrumented flux return performance: Lessons learned. *Nucl. Instrum. Methods Phys. Res. A,* 494, pp. 455–463.

Aprile, E. *et al.* (2004). Proportional light in a dual-phase xenon chamber. *IEEE Trans. Nucl. Sci.*, NS-51, p. 1986.

Arnold, R. *et al.* (1988). A ring imaging Cherenkov detector, the DELPHI Barrel RICH Prototype. Part A: Experimental studies of the detection efficiency and the spatial resolution. *Nucl. Instrum. Methods Phys. Res. A,* 270, pp. 255–288.

Attié, D. *et al.* (2021). Current status and future developments of Micromegas detectors for physics and applications. *Appl. Sci.*, 11, p. 5362.

Aubert, J. J. *et al.* (1971). Large-area wire spark chambers with two-coordinate read-out and ferrite core memory. *Nucl. Instrum. Methods.* 91, pp. 493–498.

Badertscher, A. *et al.* (2008). Construction and operation of a double phase LAr large electron multiplier time projection chamber. *IEEE Nuclear Science Symposium Conference Record.* pp. 1328–1334.

Badertscher, A. *et al.* (2011). First operation of a double phase LAr Large Electron Multiplier Time Projection Chamber with a 2D projective readout anode. *Nucl. Instrum. Methods Phys. Res. A,* 641, pp. 48–57.

Baldini, L. *et al.* (2021). Design, Construction, and Test of the Gas Pixel Detectors for the IXPE Mission. *Astropart. Phys.*, 133, p. 102628.

Balla, A. *et al.* (2013). Construction and test of the cylindrical-GEM detectors for the KLOE-2 Inner Tracker. *Nucl. Instrum. Methods Phys. Res. A,* 732, pp. 221–224.

Barney, D. (2020). Calorimetry in Particle Physics, and the CMS High-Granularity Calorimeter. *J. Instrum.*, 15, p. C07018.

Battistoni, G. *et al.* (1979). Operation of limited streamer tubes. *Nucl. Instrum. Methods,* 164, p. 57.

Battistoni, G. *et al.* (1986). The NUSEX Detector. *Nucl. Instrum. Methods Phys. Res. A,* 245, pp. 277–290.

Baussgessner, P. *et al.* (1987). The UA1 Central Detector at High Luminosity (ACOL). *Nucl. Instrum. Methods,* A257, p. 552.

Beiser, A. (1952). Nuclear emulsion technique. *Rev. Mod. Phys.*, 24(34), pp. 273–311.

Bellazzini, R. *et al.* (1999). The micro-groove detector. *Nucl. Instrum. Methods Phys. Res. A,* 424, pp. 444–458.

Bellazzini, R. *et al.* (2003). A photoelectric polarimeter based on a Micropattern Gas Detector for X-ray astronomy. *Nucl. Instrum. Methods Phys. Res. A,* 510, pp. 176–184.

Bellazzini, R. *et al.* (2004). Reading a GEM with a VLSI pixel ASIC used as direct charge collecting anode. *Nucl. Instrum. Methods*, A535, p. 477.

Bellazzini, R. *et al.* (2013). Photoelectric X-ray Polarimetry with Gas Pixel Detectors. *Nucl. Instrum. Methods Phys. Res. A,* 720, pp. 173–177.

Bellwied, R. *et al.* (2003). The STAR silicon vertex tracker: A large area silicon drift detector. *Nucl. Instrum. Methods Phys. Res. A,* 499, pp. 640–651.

Bencivenni, G. *et al.* (2017). The μ-RWELL detector. *J. Instrum.*, 12, p. C06027.

Bencivenni, G. *et al.* (2020). High space resolution μ-RWELL for high particle rate. *Nucl. Instrum. Methods Phys. Res. A,* 958, p. 162050.

Berger, M. J. *et al.* (2011). Stopping-power and range tables for electrons, protons, and helium ions.

Bernet, C. *et al.* (2005). The 40x40 cm2 gaseous microstrip detector Micromegas for the high-luminosity COMPASS experiment at CERN. *Nucl. Instrum. Methods Phys. Res. A,* 536, pp. 61–69.

Bettelli, L., Bianchi-Streit, M. and Giacomelli, G. (1993). Particle physics with bubble chamber photographs. https://cds.cern.ch/record/251912/files/cer-000168391.pdf.

Bianco, M. (2020). Upgrade of the CMS Muon Spectrometer in the forward region with the GEM technology. *J. Instrum.*, 15, p. C09045.

Bichsel, H. (1995). Calculated Bragg curves for ionization chambers of different shapes. *Med. Phys.*, 22, pp. 1721–1725.

Bisogni, M. G., Del Guerra, A. and Belcari, N. (2019). Medical applications of silicon photomultipliers. *Nucl. Instrum. Methods Phys. Res. A,* 926 (August 2018), pp. 118–128.

Boldyrev, A. S. *et al.* (2016). Tracking performance of GasPixel detectors in test beam studies. *Nucl. Instrum. Methods Phys. Res. A,* 807, pp. 47–55.

Bonivento, W. M. and Terranova, F. (2024). The science and technology of liquid argon detectors. *Rev. Mod. Phys.*, 96, p. 045001.

Bortfeldt, J. *et al.* (2018). PICOSEC: Charged particle timing at sub-25 picosecond precision with a Micromegas based detector. *Nucl. Instrum. Methods Phys. Res. A*, 903, pp. 317–325.

Bouclier, R. *et al.* (1974). Proportional chambers for a 50,000-wire detector. *Nucl. Instrum. Methods*, 115, pp. 235–244. 118–128.

Bouclier, R. *et al.* (1974). Proportional chambers for a 50,000-wire detector. *Nucl. Instr. Meth.*, 115, pp. 235–244.

Bradamante, F. and Sauli, F. (1967). *Nucl. Instr. Meth.* 56, pp. 268–276.

Brigida, M. *et al.* (2005). Investigation of the transition radiation produced by fast electrons crossing multifoil and fiber radiators. *Nucl. Instr. Meth. Phys. Res. A*, 550, pp. 157–168.

Brown, R. *et al.* (1949). Observations with electron-sensitive plates exposed to cosmic radiation. *Nature*, 183, pp. 82–87.

Brown, R. M. (2007). The CMS electromagnetic Calorimeter. *Nucl. Instr. Meth. Phys. Res. A*, 572, pp. 29–32.

Brunbauer, F. *et al.* (2018). Radiation imaging with optically read out GEM-based detectors. *JINST*, 13, p. T02006.

Brunbauer, F. M. *et al.* (2020). Radiation imaging with glass Micromegas. *Nucl. Instr. Meth. Phys. Res. A*, 955, p. 163320.

Büttner, C. *et al.* (1998). Progress with the gas electron multiplier. *Nucl. Instr. Meth. Phys. Res. A*, 409, pp. 79–83.

Buzhan, P. *et al.* (2003). Silicon photomultiplier and its possible applications. *Nucl. Instr. Meth. Phys. Res. A*, 504, pp. 48–52.

Caccia, M. *et al.* (1987). A Si strip detector with integrated coupling capacitors. *Nucl. Instr. Meth. Phys. Res. A*, 260(1), pp. 124–131.

Camarri, P. (2007). Performance of the resistive plate chambers in the ARGO-YBJ experiment. *Nucl. Instr. Meth. Phys. Res. A*, 572, pp. 476–478.

Cao, Z. (2021). Eas arrays at high altitudes start the era of uhe γ-ray astronomy. *Universe*, 7(9), pp. 1–9.

Casella, C. *et al.* (2015). A high resolution TOF-PET concept with axial geometry and digital SiPM readout. *Nucl. Instr. Meth. Phys. Res. A*, 736, pp. 161–168.

Cavanna, F., Ereditato, A. and Fleming, B. T. (2018). Advances in liquid argon detectors. *Nucl. Instr. Meth. Phys. Res. A*, 907, pp. 1–8.

Cerron Zeballos, E. *et al.* (1996). A new type of resistive plate chamber: The multigap RPC. *Nucl. Instr. Meth. Phys. Res. A*, 374, pp. 132–135.

Chardonnet, E. (2020). The DUNE dual-phase liquid argon TPC. *JINST*, 15, p. C05064.

Charpak, G. *et al.* (1968). The use of multiwire proportional counters to select and localize charged particles. *Nucl. Instr. Meth.*, 62, pp. 262–268.

Charpak, G. (1970). Evolution of the automatic spark chambers. *Ann. Rev. Nucl. Part. Sci.*, 20, pp. 196–250.

Charpak, G. *et al.* (1979). Detection of far-ultraviolet photons with the multistep avalanche chamber. Application to Cherenkov light imaging and to some problems in high-energy physics. *Nucl. Instr. Meth.*, 164, pp. 419–433.

Charpak, G. and Sauli, F. (1978). The multistep avalanche chamber: A new high rate, high accuracy gaseous detector. *Phys. Lett.*, 78 B, pp. 523–528.

Charpak, G., Sauli, F. and Duinker, W. (1973). High-accuracy drift chambers and their use ib strong magnetic fields. *Nucl. Instr. Meth.*, 108, pp. 413–426.

Chechik, R. *et al.* (2006). Thick GEM-like (THGEM) detectors and their possible applications. *SNIC Symposium Stanford.*

Cherkashyna, N. *et al.* (2015). Overcoming high energy backgrounds at pulsed spallation sources. *arXiv:1501.02364v1.*

Chikovani, G. E., Mikhailov, V. A. and Roinishvili, V. N. (1963). A track spark chamber. *Phys. Lett.*, 6(3), pp. 254–255.

Chochula, P. *et al.* (1998). The DELPHI silicon tracker at LEP2. *Nucl. Instr. Meth. Phys. Res. A*, 412, pp. 304–328.

Clergeau, J. F. *et al.* (2001). Operation of sealed microstrip gas chambers at the ILL. *Nucl. Instr. Meth. Phys. Res. A*, 471, pp. 60–68.

CMS_Collaboration (1997). The electromagnetic calorimeter technical design report CERN/LHCC 97-033. *CMS TDR 4*.

CMS_Collaboration (2008). The CMS experiment at the CERN LHC. *JINST*, 3, p. S08004.

Contin, G. (2017). MAPS-based vertex detectors: Operational experience in STAR and future applications. In *VERTEX 2017*.

Cozza, D. *et al.* (2003). The CSI-based RICH detector array for the identification of high momentum particles in ALICE. *Nucl. Instr. Meth. Phys. Res. A*, 502, pp. 101–107.

Curceanu, C. *et al.* (2019). X-ray detectors for kaonic atoms research at DAΦNE. *Condens. Matter*, 4(2), pp. 1–14.

Damerell, C. (2021). State-of-the-art tracking for high luminosities. *CERN Courier*, 61(4), pp. 37–46.

D'Ambrosio, C. *et al.* (1993). Imaging performance obtained with fused bundles of scintillating fibres. *Nucl. Instr. Meth. Phys. Res. A*, 325(1–2), pp. 161–167.

D'Ambrosio, C. *et al.* (1994). Photon counting with a hybrid photomultiplier tube (HPMT). *Nucl. Instr. Meth. Phys. Res. A*, 338(2–3), pp. 389–397.

Daniel, R. R. *et al.* (1952). Evidence for the Creation of Heavy Mesons. *Phil. Mag. and J. of Science*, 43(342), pp. 753–774.

Dalla Betta, G. F. and Povoli, M. (2022). Progress in 3D Silicon radiation detectors. *Front. Phys.*, 10(June), pp. 1–17.

Damerell, C. (1998). Charge-coupled devices as particle tracking detectors. *Rev. Sci. Instrum.*, 69(4), pp. 1549–1573.

Damerell, C. (2023). Tracking the rise of pixel detectors. *CERN Courier*, pp. 1–17.

Da Via, C. *et al.* (2012). 3D silicon sensors: Design, large area production and quality assurance for the ATLAS IBL pixel detector upgrade. *Nucl. Instr. Meth. Phys. Res. A*, 694, pp. 321–330.

Da Via, C. (2012). Silicon sensors go 3D. *CERN Courier*, (May), pp. 1–6.

Decamp, D. *et al.* (1990). ALEPH: A detector for electron-positron annihilations at LEP. *Nucl. Instr. Meth.*, A294, p. 121.

Di Sciascio, G. (2014). Main physics results of the ARGO-YBJ experiment. *Int. J. Mod. Phys. D*, 23, p. 1430019.

Dolgoshein, B. (1993). Transition radiation detectors. *Nucl. Instr. Meth.*, A326, p. 434.

Dolgoshein, B. A., Rodionov, B. U. and Luchkov, B. I. (1957). Streamer Chamber. *Nucl. Instr. Meth.*, 29, pp. 270–276.

Drumm, H. *et al.* (1980). Experience with the Jet-Chamber of the JADE detector at PETRA. *Nucl. Instr. Meth.*, 176, p. 333.

Evans, H. G. (1991). Construction and filling of a uranium tetramethyl-pentane calorimeter. *Nucl. Phys. B Proc. Suppl.*, 23(1), pp. 80–91.

Fabjan, C. W. and Gianotti, F. (2003). Calorimetry for particle physics. *Rev. Mod. Phys.*, 75, p. 1243.

Fancher, D. *et al.* (1979). Performance of a time-projection chamber. *Nucl. Instr. Meth.*, 161(3), pp. 383–390.

Fich, O. *et al.* (1975). Channeling of 1.1 GeV/c protons and pions. *Phys. Lett. B*, 57(1), p. 90.

Fiorini, M. (2020). The upgrade of the LHCb RICH detectors. *Nucl. Instr. Meth. Phys. Res. A*, 952, p. 161688.

Fonte, P. (2013). Survey of physical modelling in Resistive Plate Chambers. *J. Instrum.*, 8, p. P11001.

Fonte, P., Smirnitski, A. and Williams, M. S. C. (2000). A new high-resolution TOF technology. *Nucl. Instr. Meth.*, A 443, p. 201.

Fraga, F. A. F. *et al.* (2001). Optical readout of GEMs. *Nucl. Instr. Meth. Phys. Res. A*, 471, pp. 125–130.

Fraga, F. A. F. *et al.* (2002). CCD readout of GEM-based neutron detectors. *Nucl. Instr. Meth. Phys. Res. A*, 478, pp. 357–361.

French, M. J. *et al.* (2001). Design and results from the APV25, a deep sub-micron CMOS front-end chip for the CMS tracker. *Nucl. Instr. Meth.*, A 466, p. 359.

Fukuda, S. *et al.* (2003). The Super-Kamiokande detector. *Nucl. Instr. Meth. Phys. Res. A*, 501, pp. 418–462.

Fukui, S. and Miyamoto, S. (1959). A new type of particle detector: The «discharge charmber». *Nuovo Cimento Ser. 10*, 11(1), pp. 113–115.

Gatti, E. and Rehak, P. (1984). Semiconductor drift chamber - An application of a novel charge transport scheme. *Nucl. Instr. Meth. Phys. Res.*, 225(3), pp. 608–614.

Gilmore, R. S. (1980). Particle Identification by Cherenkov and Transition Radiation. *SLAC-PUB-2506*.

Giomataris, I. *et al.* (2006). Micromegas in a bulk. *Nucl. Instr. Meth. Phys. Res. A*, 560, pp. 405–408.

Giomataris, Y. *et al.* (1996). MICROMEGAS: A high-granularity position-sensitive gaseous detector for high particle-flux environments. *Nucl. Instr. Meth. Phys. Res. A*, 376, pp. 29–35.

González-Díaz, D., Monrabal, F. and Murphy, S. (2018). Gaseous and dual-phase time projection chambers for imaging rare processes. *Nucl. Instr. Meth. Phys. Res. A*, 878, pp. 200–255.

van der Graaf, H. (2007). GridPix: An integrated readout system for gaseous detectors with a pixel chip as anode. *Nucl. Instr. Meth. Phys. Res. A*, 580, pp. 1023–1026.

van Der Graaf, H. (2010). Gaseous detectors. *Nucl. Instr. Meth. Phys. Res. A*, 628, pp. 27–30.

Gys, T. (1999). An overview of current developments in position-sensitive hybrid photon detectors and photo-multiplier tubes. *IEEE Nucl. Sci. Symp. Conf. Rec.*, pp. 39–45.

Hamamatsu (2017). *Photomultiplier Tubes: Basics and Applications*.

Hauer, P. (2022). The upgraded ALICE TPC. *Nucl. Instr. Meth. Phys. Res. A*, 1039, p. 167023.

Heijne, E. *et al.* (1980). A silicon surface barrier microstrip detector designed for high energy physics. *Nucl. Instr. Meth.*, 178(2–3), pp. 331–343.

Heijne, E. H. M. (2021). History and future of radiation imaging with single quantum processing pixel detectors. *Radiat. Meas.*, 140, p. 106436.

Henderson, C. (1970). *Cloud and bubble chambers*. Methuen's Monographs.

Herz, A. J. and Lock, W. O. (1966). The particle detectors 1. Nuclear emulsions. In *The Particle Detectors CERN*, pp. 83–87.

Hubbell, J. H. and Seltzer, S. M. (1996). X-ray mass attenuation coefficients.

Iarocci, E. (1983). Plastic streamer tubes and their applications in high energy physics. *Nucl. Instr. Meth.*, 217, p. 30.

Ieva, M. (2010). The European Scanning System for the measurement and analysis of OPERA neutrino interactions with nuclear emulsions. *Nucl. Instr. Meth. Phys. Res. A*, 617, pp. 275–279.

Inami, K. *et al.* (2006). A 5 ps TOF-counter with an MCP-PMT. *Nucl. Instr. Meth. Phys. Res. A*, 560, pp. 303–308.

Ingram, Q. *et al.* (2001). Avalanche photodiodes for the CMS electromagnetic calorimeter. *Proc. Sci.*, 7(July).

Järvinen, M. L. and Sipilä, H. (1982). Effect of pressure and admixture of neon penning mixtures on proporional counter resolution. *Nucl. Instr. Meth.*, 193, pp. 37–38.

Joram, C., Haefeli, G. and Leverington, B. (2015). Scintillating fibre tracking at high luminosity colliders. *J. Instrum.*, 10(8).

Kemmer, J. (1980). Fabrication of low noise silicon radiation detectors by the planar process. *Nucl. Instr. Meth.*, 169(3), pp. 499–502.

Klempt, W. (1999). Review of particle identification by time of flight techniques. *Nucl. Instr. Meth. Phys. Res. A*, 433, pp. 542–553.

Koppert, W. J. C. *et al.* (2013). GridPix detectors: Production and beam test results. *Nucl. Instr. Meth. Phys. Res. A*, 732, pp. 245–249.

Krammer, M. (2009). Silicon Sensors: From basic principles to the largest Silicon detector. In *IPM School*, pp. 117–126.

Krienen, F. (1962). Digitized spark chambers. *Nucl. Instr. Meth.*, 16, p. 262.

Kutter, T. (2023). A novel highly segmented neutrino detector: The super fine grained detector for the upgraded T2K ND280 near detector. *Proc. Sci.*, (July 2024).

Ligtenberg, C. *et al.* (2020). The gaseous QUAD pixel detector. *Nucl. Instr. Meth. Phys. Res. A*, 958, p. 162731.

Lippmann, C. (2012). Particle identification. *Nucl. Instr. Meth. Phys. Res. A*, 666, pp. 148–172.

Litt, J. and Meunier, R. (1973). Cherenkov counters technique in high-energy physics. *Ann. Rev. Nucl. Sci.*, 23, pp. 1–44.

Liu, Z. *et al.* (2018). 20 gas gaps Multigap Resistive Plate Chamber: Improved rate capability with excellent time resolution. *Nucl. Instr. Meth. Phys. Res. A*, 908, pp. 383–387.

Llopart, X. *et al.* (2002). Medipix2: A 64-k pixel readout chip with 55-μm square elements working in single photon counting mode. *IEEE Trans. Nucl. Sci.*, NS-49, p. 2279.

Llopart, X. *et al.* (2007). Timepix, a 65k programmable pixel readout chip for arrival time, energy and/or photon counting measurements. *Nucl. Instr. Meth. Phys. Res. A*, 581, pp. 485–494.

Llope, W. J. (2005). The large-area time-of-flight upgrade for STAR. *Nucl. Instr. Meth. Phys. Res. B*, 241, pp. 306–310.

Marco-Hernández, R. (2020). Overview of cmos sensors for future tracking detectors. *Instruments*, 4(4), pp. 1–9.

Marel, G. *et al.* (1977). Large planar drift chambers. *Nucl. Instr. Meth.*, 141, pp. 43–56.

McCarthy, R. *et al.* (1986). Identification of large-transverse-momentum hadrons using a ring-imaging Cherenkov counter. *Nucl. Instr. Meth. Phys. Res. A*, 248, pp. 69–75.

Meunier, R. *et al.* (1962). DISC: Compteur Cherenkov a selection de Vitesse. *Nucl. Instr. Meth.*, 17, pp. 1–19.

Mindur, B. (2017). ATLAS transition radiation tracker (TRT): Straw tubes for tracking and particle identification at the Large Hadron Collider. *Nucl. Instr. Meth. Phys. Res. A*, 845, pp. 257–261.

Neutron Cross Section of the elements (2017).

Nygren, D. and Marx, J. (1978). The time projection chamber. *Phys. Today*, 31, p. 46.

Nygren, D. R. (2018). Origin and development of the TPC idea. *Nucl. Instr. Meth. Phys. Res. A*, 907, pp. 22–30.

Oed, A. (1988). Position-sensitive detector with microstrip anode for electron multiplication with gases. *Nucl. Instr. Meth. Phys. Res. A*, 263, pp. 351–359.

OPAL_Collaboration (1991). The OPAL detector at LEP. *Nucl. Instr. Meth. Phys. Res. A*, 305(2), pp. 275–319.

Otte, A. N. *et al.* (2005). A test of silicon photomultipliers as readout for PET. *Nucl. Instr. Meth. Phys. Res. A*, 545, pp. 705–715.

Pansky, A. *et al.* (1995). The scintillation of CF, and its relevance to detection science. *Nucl. Instr. Meth. Phys. Res. A*, 354, pp. 262–269.

Paolucci, L. (2022). The LHCb muon detector upgrades. *Proc. Sci.*, 414, pp. 0–5.

Parker, S. I., Kenney, C. J. and Segal, J. (1997). 3D - A proposed new architecture for solid-state radiation detectors. *Nucl. Instr. Meth. Phys. Res. A*, 395, pp. 328–343.

Periale, L. *et al.* (2002). Detection of the primary scintillation light from dense Ar, Kr and Xe with novel photosensitive gaseous detectors. *Nucl. Instr. Meth. Phys. Res. A*, 478, pp. 377–383.

Picture CERN (as for many others)

Piemonte, C. and Gola, A. (2019). Overview on the main parameters and technology of modern Silicon Photomultipliers. *Nucl. Instr. Meth. Phys. Res. A*, 926, pp. 2–15.

PNSensor GmbH (2018). https://www.pnsensor.de/.

Policarpo, A. J. P. L. *et al.* (1972). Improved resolution for low energies with gas proportional scintillation counters. *Nucl. Instr. Meth.*, 102, pp. 337–348.

Powell, C. F. *et al.* (1960). The study of elementary particles by the photographic method. *Phys. Today*, 9, p. 45.

Pühlhofer, F., Röhrich, D. and Keidel, R. (1988). Track recognition in digitized streamer chamber pictures. *Nucl. Instr. Meth. Phys. Res. A*, 263(2–3), pp. 360–367.

Keim, R. (2020). *Understanding the Structure and Functionality of CCDs.*

Ratcliff, B. and Va'vra, J. (2020). DIRC: Internally reflecting imaging Cherenkov detectors. *Nucl. Instr. Meth. Phys. Res. A*, 970, p. 163442.

Riedler, P. (2015). Overview of monolithic silicon pixel detectors. In *CLIC Seminar.*

Roberts, A. (1992). The birth of high-energy neutrino astronomy: A personal history of the DUMAND project. *Rev. Mod. Phys.*, 64, pp. 259–312.

Rohrbach, F. (1988). *Streamer Chambers at CERN during the past decade and visual techniques of the future*. CERN/EF 88-17.

Rubbia, C. *et al.* (2011). Underground operation of the ICARUS T600 LAr-TPC: First results. *JINST*, 6, p. P07011.

Rutherford, E. and Geiger, H. (1908). An electrical method of counting the number of alpha-particles from radio-active substances. *Proc. R. Soc. A*, 81, p. 141.

Saito, K. *et al.* (2002). Absolute number of scintillation photons emitted by alpha particles in rare gases. *IEEE Trans. Nucl. Sci.*, 49, pp. 1674–1680.

Santonico, R. and Cardarelli, R. (1981). Development of resistive plate counters. *Nucl. Instr. Meth.*, 187, pp. 377–380.

Sauli, F. (1997). GEM: A new concept for electron amplification in gas detectors. *Nucl. Instr. Meth. Phys. Res. A*, 386, pp. 531–534.

Sauli, F. (2016). The gas electron multiplier (GEM): Operating principles and applications. *Nucl. Instr. Meth. Phys. Res. A*, 805, pp. 2–24.

Schilly, P. *et al.* (1970). Construction and performance of large multiwire proportional chambers. *Nucl. Instr. Meth.*, 91, p. 221.

Schyns, E. (2002). Status of large area CsI photocathode developments. *Nucl. Instr. Meth.*, A494, p. 441.

Scintillator Library (2024).

Sekiya, M. and Yamasaki, M. (2015). Antoine Henri Becquerel (1852–1908): A scientist who endeavored to discover natural radioactivity. *Radiol. Phys. Technol.*, 8, pp. 1–3.

Selove, W., Kononenko, W. and Wilsker, B. (1979). Fluorescent-wave-shifters and scintillators for calorimeters. *Nucl. Instr. Meth.*, 161(2), pp. 233–242.

Sguazzoni, G. (2008). The construction and commissioning of the CMS Silicon Strip Tracker. In *34 Int Conf on High Energy Physics*.

Smirnov, I. B. (2005). Modeling of ionization produced by fast charged particles in gases. *Nucl. Instr. Meth. Phys. Res. A*, 554, pp. 474–493.

Suzuki, M. *et al.* (1988). On the optical readout of gas avalanche chambers and its applications. *Nucl. Instr. Meth. Phys. Res. A*, 263, pp. 237–242.

Terzo, S. *et al.* (2021). Novel 3D pixel sensors for the upgrade of the ATLAS inner tracker. *Front. Phys.*, 9(April), pp. 1–15.

Turala, M. (2005). Silicon tracking detectors - Historical overview. *Nucl. Instr. Meth. Phys. Res. A*, 541, pp. 1–14.

Utrobicic, A. *et al.* (2023). A large area 100 channel Picosec Micromegas detector with sub 20 ps time resolution. *MPG2022 Subm JINST*.

Va'vra, J. (1999). Long-term operational experience with the barrel CRID at SLD. *Nucl. Instr. Meth. Phys. Res. A*, 433, pp. 59–70.

Veloso, J. F. C. A. *et al.* (2010). Energy resolved X-ray fluorescence imaging based on a micropattern gas detector. *Spectrochim. Acta Part B*, 65, pp. 241–247.

Walenta, A. H. *et al.* (1971). The multiwire drift chamber: A new type of proportional wire chamber. *Nucl. Instr. Meth.*, 92, pp. 373–380.

Walenta, A. H. (1979). The time expansion chamber and single ionization cluster mesurement. *IEEE Trans. Nucl. Sci.*, NS-26, p. 73.

Wang, Y. and Yu, Y. (2021). Multigap resistive plate chambers for time of flight applications. *Appl. Sci.*, 11, pp. 1–23.

Wilkens, H. (2009). The ATLAS liquid argon calorimeter: An overview. *J. Phys. Conf. Ser.*, 160, p. 012043.

Williams, M. C. S. (2012). Particle identification using time of flight. *J. Phys. G Nucl. Part. Phys.*, 39(12).

Wilson, J. G. (2014). *The principles of Cloud-Chamber Technique*. Cambridge University Press.

Workman, R. L. *et al.* (2022). Review of particle physics. *Prog. Theor. Exp. Phys.*, 8.

Ypsilantis, T. (1981). Cerenkov ring imaging. *Phys. Scr.*, 23, pp. 371–376.

Yu, S. *et al.* (2023). Recent neutrino oscillation results with the IceCube experiment. *arXiv:2307.15855*, p. 1143.

Index

www.ingramcontent.com/pod-product-compliance
Ingram Content Group UK Ltd.
Pitfield, Milton Keynes, MK11 3LW, UK
UKHW022103170926
8624IPUK00024B/38

9 789819 819393